Ecology Crisis

God's Creation and Man's Pollution

John W. Klotz

Concordia Publishing House

St. Louis London

Concordia Publishing House, St. Louis, Missouri
Concordia Publishing House Ltd., London, E. C. 1

Library of Congress Catalog Card No. 70-150211
ISBN 0-570-03117-6

MANUFACTURED IN THE UNITED STATES OF AMERICA

Contents

Chapter 1

An Overview of the Problem

Problems of environmental deterioration have been with us for a long time. A. D. 61 the Roman poet Seneca wrote "As soon as I had gotten out of the heavy air of Rome and from the stink of the smoky chimneys thereof, which, being stirred, poured forth whatever pestilent vapors and soot they held enclosed in them, I felt an alteration of my disposition." In 1257 Eleanor of Aquitane, the queen of Henry II, moved from Nottingham to Tutbury Castle to get away from what she called "the undesirable smoke," and in 1661 the diarist John Evelyn described London as follows "And what is all this but the hellist and dismal cloud of sea coal which is not only perpetually imminent but so universally mixed with the otherwise wholesome and excellent air that London's inhabitants breathe nothing but impure and thick mist accompanied with a filthy vapor which renders them susceptible to thousand inconveniences corrupting the lungs and disordering the entire habit of their body so that catarrhs, coughs, and consumptions rage more in this city than in the whole earth besides."

Samuel Taylor Coleridge was so moved by what he observed on a 19th century visit to Cologne that he wrote:

In Köln, a town of monks and bones,
And pavements fanged with murderous stones,
And rags and hags and hideous wenches,
I counted two and seventy stenches,
All well defined, and several stinks.
Ye nymphs that reign o'er sewers and sinks,
The River Rhine as is well known,
Doth wash your city of Cologne;
But tell me nymphs, what power divine,
Shall henceforth wash the River Rhine?

Environmental Deterioration in the Past

Actually, damage to our soils antedates damage to the air and water. The earliest civilization developed in the fertile crescent from the valley of the Tigris and the Euphrates to the land of Egypt. Today the area watered by the Tigris and the Euphrates, where the ancient Sumerian, Babylonian, and Syrian civilizations thrived, is desert. The headwaters of the Tigris and the Euphrates lie in the highlands of Armenia in areas that in the past have supported high populations of people and higher numbers of sheep and goats. These headwater areas have been subjected to all the pressures which hillside farming and overgrazing by livestock can cause. In addition, the forest which held the soil has been removed to provide building materials for the growing cities or to supply additional grazing land for flocks and herds.

The erosion that resulted from the removal of cover brought about an ever-increasing silt load to be carried by the Tigris and the Euphrates. In Sumerian times it is generally believed that the silt load was manageable, but subsequent empires found it increasingly difficult to control. Armies of laborers and slaves were kept busy removing the silt load from the irrigation canals. The silt reached the Persian Gulf and filled it in for as much as 180 miles so that what were once harbors are now that distance inland. With the decline of agriculture due to silting in of the irrigation canals, the land was unable any longer to support the populations it once sustained, and commerce disappeared with the filling in of harbors. The disrepair of the irrigation canals made agriculture impossible.

Lebanon is another country which shows what happens when the land is misused. The Phoenicians founded their maritime empire and built the greatest navy of their day with the timber that grew on the mountains. These were the famous cedars of Lebanon that helped shape the Egyptian cities and were even used in the temple of Solomon. It was the cutting of the timber which began the trouble. Regeneration of the forest was impossible because the cleared lands were heavily grazed by sheep and goats. Many of the formerly forested hills are now completely barren and almost devoid of topsoil. From their appearance it might be thought that the climate is too dry to support trees; yet where soil remains in a few protected spots there are ancient groves, and here the cedars continue to reproduce and grow.

Egypt has been more fortunate. Because of the nature of the annual floods and because the sources of the Nile are relatively undisturbed, the valley continues to be as fertile or almost as fertile as in ancient times. One author has described Egypt as "manproof." It may be, though, that this country will be affected by the changes man is making in building the Aswan Dam—we shall refer to these later.

In the New World, in the early centuries of our era, the Mayas built great cities with huge pyramids, massive masonry, and elaborate carvings. Food was there in abundance, and fields were easily cleared by girdling trees with sharp stone hatchets and burning the timber. Yet by the 6th century the Central American cities were abandoned, and another civilization was established farther to the north. The people were not exterminated nor conquered, there is no evidence of disease serious enough to permanently wipe out a civilization, and it is quite likely that their agricultural methods destroyed the fertility of the tropical soils, easily leached of chemical elements, so that the civilization could not continue.

In our own country the settlers found a rich fertile land. The streams ran clear, and there were few floods. It was not long until our fathers changed all that. The streams filled with silt and mud, the fish died, the land eroded and became scarred with gullies—and our fathers moved west to exploit the virgin lands there. Early Massachusetts records indicate that most of the land near the coast was abandoned at least once before 1800.

Modern Concern for the Environment

Concern for the environment began in Europe long before it began in the United States. Its basis there was often a selfish one. Royalty and nobility desired undisturbed areas in which they could hunt, and these were protected against the inroads both of agriculture and poachers.

Concern Begins in the United States

Concern for the environment in the United States began around the turn of the century with the extinction of the passenger pigeon and the near extinction of the bison. President Theodore Roosevelt was a leader in this movement. In May 1908 he called a conference of governors at the White House. This was the first conference to which governors of all states were invited for consideration of any question of national policy. Leading up to the conference were a number of events adversely affecting the environment and the enactment and carrying out of a series of laws governing the disposal of the public domain. These included the spectacular lumbering operations which had destroyed much of the timber of the lake states from 1870 on; rapid growth of population from 1890 on with the prediction that the United States would have more than 200 million people in 1950, and a growing concern that the country was not managing its resources wisely to take care of all the people; the Homestead Act of 1862; the mining laws and the Timber and Stone Act of 1878; consideration of forest exploitation by the American Association for the Advancement of Science and the presentation of memorials to Congress in 1870 and again in 1890, which resulted in the establishment of a Forestry Service in the U. S. Department of Agriculture in 1890 and in the legalization and withdrawal from settlement of the first forest reserves in 1891; the studies and publication of a report by Major J. W. Powell on "Lands of the Arid Region" in 1879, and the establishment of an irrigation division in the U. S. Geological Survey; recommendations by the

National Academy of Sciences in 1897 that the forest reserve policy be strengthened; long agitation for and the passage of the Reclamation Act in 1902; and the appointment of the Inland Waterways Commission by President Roosevelt in 1907, the first report of which emphasized the interrelated character of water, forest, transportation, and fuel.

The White House conference produced a "Declaration of Principles" which suggested that some order was needed in the use and management of natural resources. Two men left their mark on the conference: President Roosevelt and Gifford Pinchot, chief of the U. S. Forestry Service and chairman of the Inland Waterways Commission. As a result of the conference a 50-man national conservation commission composed about equally of scientists, businessmen, and statesmen was appointed to report back to the governors in December 1908 with an inventory of the country's natural resources. Subsequently the governors of 41 states appointed state conservation agencies.

Later, in February 1909 President Roosevelt called a conference of North American nations to meet in Washington to consider natural resource questions on a hemispheric scale.

Conservation Bogs Down

Unfortunately, following this fine beginning some of the work of early conservationists became based more on enthusiasm than on facts. Gifford Pinchot said in 1910, "We have timber for less than 30 years, anthracite coal for about 50 years. . . . Supplies of iron ore, minerals, oil, and natural gas are being rapidly depleted." Many people were unimpressed with the expressed concerns of the early conservationists when predictions such as those of Pinchot, a recognized authority, did not come true. The feeling developed that our resources were indeed endless and that damage to the environment could be ignored with impunity.

As a result people ignored the damage that was being done to their surroundings and even regarded evidence of it as proof for an advanced technology. The city of Pittsburgh prided itself on being "the Smoky City," and residents there were concerned when they saw blue skies. The heavy overcast indicated that the mills were operating at full capacity and that the economy was healthy. Blue sky meant a downturn in the economy and unemployment.

The great depression proved to be a blessing in disguise so far as the environment was concerned, for President Franklin D. Roosevelt found the need for a great public works program to furnish employment, and this he used to protect and develop our natural resources. The Civilian Conservation Corps accomplished a great deal in forest protection, forest planting and care, soil erosion control, pest control, lake and stream improvement, recreational development, and flood control. In 1935 the Soil Conservation Service, set up as a result of the dust bowl emergency, was established in the Department of Agriculture. The Tennessee Valley Authority established in 1933 also was involved in protecting the environment.

Chapter 2

Our Deteriorating Environment

Real awareness of the seriousness of our problems and of the magnitude of environmental deterioration has come only within the last few decades. Our concern has focused particularly on problems of atmospheric pollution. One of the first of these was the famous Donora smog which began on Tuesday, October 26, 1948. Smogs were so common in Donora that this was more or less a way of life. The city is located in a valley of the Monongahela River on a rather wide bend, with bluffs rising some 450 feet above the town. It's a typical Pennsylvania mill town with mills in the center of town and abandoned coal mines on the hills surrounding it. About two thirds of the people who live in town are employed in the mills. Donora had a population of about 12,500 when the killer smog struck.

By Thursday, October 28, visibility was practically zero, and the sickening smell and taste of sulphur dioxide were everywhere. By Friday it was noted that smoke from locomotives didn't seem to rise: it just spilled over the smokestacks, and by late that afternoon asthmatics were beginning to show up in doctors' offices.

The Crisis Begins

One physician reports that his phone started to ring at 4 o'clock Friday afternoon and did not stop until the smog lifted. By 6 o'clock all of the doctors in town realized that something unusual was up. Most of them reported that they had time only to return to their offices for drugs. The smog not only created a medical emergency, but it also made getting around in the town difficult. It was impossible to drive, and one doctor even reported that he had trouble finding his office when he sought to return for additional drugs.

By evening the fire department began to get inhalator calls. Motorists reported that their autos would stall when they took their foot off the accelerator.

The first death occurred at 1:30 a. m. on Saturday morning. The undertaker who received the call was only two blocks from the home in which the death occurred, but it took him 30 minutes to reach the scene and return with the body. No sooner had he returned to his mortuary than the phone rang with another case—something very unusual in a town of only 12,500. This particular call was from several miles out in the country, and he found it necessary to get out of the hearse and walk ahead of the driver so that the hearse would not leave the road.

By 10 o'clock the next morning there were 10 bodies in his mortuary.

An emergency first aid station was opened the next evening, but suddenly the calls slackened as a gentle rain dissipated the smog. However, by that time there were 22 dead and 5,910 so ill that they sought medical help either from the town physicians, from the fire department inhalator squad, or from the emergency first aid station that had been set up.

Later studies indicated that the fog was caused by an atmospheric inversion, a rather common condition where pollutants are held close to the earth by a layer of warm air. Ordinarily polluted air is warmer than pure air and rises gradually into the clean air where it is diluted. However, at Donora there was a layer of warm air which served as a lid to keep all of the polluted air between the river bluffs.[1]

Other Smog Emergencies

The Donora disaster was not the first instance of fatalities due to air pollution. In 1930 the 62 deaths in Belgium's Meuse Valley were attributed to smoke and smog. Later, on December 5, 1952, a thick fog settled over London, and during the 5-day smog it is estimated that 4,000 more deaths than expected occurred. In the next two months there were 8,000 more deaths than the expected number, and these were attributed to the after-effects of the smog. In 1956 there were 1,000 deaths in London attributed to the smog and in 1962, 300. In 1953, some 200 excessive deaths in New York City were attributed to the smog there and in 1963, 400. In 1965 a smog at Thanksgiving in New York City is believed to have been responsible for 80 additional deaths.

The smogs which occurred in the Meuse Valley, Donora, London, and New York were *reducing smogs.* This smog contains sulphur dioxide which on further oxidation forms sulphuric acid, one of the most corrosive chemicals known. It also contains carbon in the form of soot and fly ash; these particles ride on various poisonous gases including benzopyrene.

The Los Angeles Smog

The famous Los Angeles smog of which we have heard much in recent years is an *oxidizing smog.* It is not composed of smoke but of gases including vapors from internal combustion engines. Nitrogen dioxide, ozone, and peroxyacetyl nitrate (PAN) are the chief offenders. This type of smog makes the nose run and the eyes water. It is believed that PAN, which is produced by the action of sunlight on the waste products from gasoline vapor, is the chief offender. Los Angeles is characterized by frequent temperature inversions, the meteorological phenomenon which was responsible for the Donora smog to which we have referred.

An Increasing Awareness of Our Problem

Because we are finally realizing the dangers of our pollution, we have been reading such statements as "garbage in the atmosphere is as unhealthy as garbage in the drinking water. We must face the fact that clean air, unlike clean water, can't be imported"—a reference to the fact that New York City imports its drinking water from Catskill reservoirs. Another writer has said "Pure air is like money. It is only important to those who do not have it" and Morris Neiburger, a Los Angeles meteorologist, pessimistically suggests that all civilization will pass away not from a sudden cataclysm like a nuclear war but from gradual suffocation in its own wastes. We have also read and heard repeated references to "our effluent society," a cynical reference to the wastes which we are producing, and so in the last third of the century we have finally come to recognize that "man has been wasteful of the resources of the world in which he lives. He has ravaged its forests and soils and has plundered its mineral wealth; he has squandered and soiled its waters; he has contaminated its air. No reasonable person would suggest that man not use his environment, or that he revert to his primitive past. But no reasonable person can condone his wasteful excesses." [2]

Why Has Deterioration Been Permitted?

Part of the problem has arisen because man has failed to recognize that he is a part of the ecosystem. The myth of man's separateness from his environment has been encouraged by the myth of scientific and technical omnipotence. The general public has been persuaded that any problem can be solved by appropriating a sufficient amount of money and by devoting the best available brains to the solution of the problem. After all, we are told, medical research has resulted in a more than doubling of the human life expectancy in the past two centuries, and in recent years we have seen how science and technology have solved all sorts of problems as they have arisen.

Scientific Triumphs

For instance, reference is made to the development of nuclear weapons. It was decided that we wanted a super bomb in order to gain victory in World War II. The most competent and the most gifted of the scientific community were recruited, they were given a blank check so far as funds were concerned, and the result was the production of the A-bomb.

Similarly it was decided that we needed a preventative for polio. Once more the best brains were recruited, and these men were given unlimited funds; the result was the Salk and Sabin vaccines which have virtually eliminated this killer and crippler of children and adults as a medical problem. Likewise we decided in 1957 that we would like to explore space, and we determined that the first men on the moon should be Americans. The best scientists were recruited, once more they were given unlimited funds, and the result was that Americans landed on the moon.

And so complacency has been encouraged by statements that all that we need to do is to appropriate the necessary funds and to place adequate resources at science's disposal in order to restore the quality which has been lost from the environment. Man exists apart from nature; science and technology are capable of solving all problems.

The Ecosystem

But the fact of the matter is that man is a part of a complex ecosystem. In dealing with the environment we are not dealing with the simple relationships which have been involved in the problems recently solved through science and technology. The development of the A-bomb, the conquest of polio, the landing of men on the moon involved solving relatively few cause and effect relationships. Problems of the environment are much more complex. They involve complicated food chains beginning with simple plants which trap and store up the energy of the sun. The simple plants may be eaten by animals, they may be eaten directly by man, or they may die and be broken down by decay organisms.

Even man cannot be said to be at the end of the chain; rather there is a matter-energy cycle with energy being released and chemical compounds simplified by decay processes so that these substances can once more be incorporated into protoplasm. These cycles and chains are very complex, they are a tribute to God's creative wisdom, and they challenge the wisest human minds when man seeks to understand and unravel them.

Autophytes and Heterophytes

All ecosystems have two basic components: the autophytic component, the plants which fix the energy of the sun by synthesizing their food from inorganic substances; and the heterophytic component which utilizes the food stored by the autophytes, rearranges it, and finally decomposes complex materials into simple organic compounds again. Thus the energy which the plants store passes through the community in a series of steps of eating and being eaten which scientists know as the food chain. Food chains are actually interlinked to form a food web so that the food web is a more accurate concept than the food chain.

A number of species consume the basic plant material. The scientist recognizes that there are a number of organisms which obtain their food in the same number of steps, and these are said to be on the same trophic level. Some organisms may occupy a single trophic level but higher organisms occupy several trophic levels.

At each step in the food chain a considerable portion of the potential energy is lost as heat; organisms at each trophic level pass on less energy than they receive. This fact, though self-evident, is very important. In most cases therefore food chains are limited to four or five levels; the longer the food chain the less energy is available for the final members.

Herbivores

The first trophic level, the green plants, are the producers, and they are the base on which the heterotrophic component rests. The next trophic level consists of the herbivores. These convert the energy stored in plant tissues into animal tissue. They are adapted to a diet that is high in cellulose and are characterized by special teeth, complicated stomachs, long intestines, a well developed caecum, and symbiotic flora and fauna – simple plants and animals that live in the intestinal tract without harming their hosts. Deer and other ruminants are a good example of this trophic level. The chief land herbivores are insects, rodents, and hoofed animals. In lakes and seas minute crustaceans are the herbivores which graze on the diatoms, the autophytes of this habitat.

Carnivores

The next trophic level consists of the carnivores. Those feeding directly on herbivores are first level carnivores or second level consumers. They are larger and stronger than their prey and more or less solitary in habit. The first level carnivores are an energy source for second level carnivores, and there may be third and fourth level carnivores. As the trophic level increases the number of organisms decreases, and their fierceness, agility, and size tends to increase.

Omnivores

Omnivores, of which man is an example, eat both plants and animals and occupy several trophic levels. The red fox, for instance, feeds on insects, small rodents, and even dead animals. The white-footed mouse is basically an herbivore, but he also eats insects, small birds, and bird eggs.

Energy Reduction

The amount of energy is reduced in magnitude by 100 from the primary producers to the plant consumers (the herbivores) so that 1,000 calories of solar energy stored up by the autophytes becomes 10 calories of herbivore tissue. The amount of energy is reduced by 10 for each step thereafter through the first, second, and third level carnivores. It is for this reason that most food chains can have only three or four links and that five is a distinct luxury.

A practical application of this principle to problems of food supplies for man is the recognition that we can support a much larger population by using plant materials than by using meat. It is apparent that as populations continue to climb more and more people will have to live on the trophic level next to the plants. Man can support even larger populations by cooking plants to make them more digestible; in that way more energy becomes available to him from plants.

Cycles

There are many complicated cycles which are a part of the ecosystem. One of these is the oxygen-carbon dioxide cycle. The green plants not only store up the sun's energy in the form of food but also release oxygen as a byproduct of this process of photosynthesis. The raw material that they use is the carbon dioxide which is the product of respiration, the process by which the sun's energy is released in the cells of all living things. We are most conscious of respiration as it occurs in animal cells, but respiration is also a part of plant metabolic processes; however, because far more oxygen is produced as a result of photosynthesis in plants than the amount of carbon dioxide produced as a result of respiration, we usually think of the green plants as oxygen producers, and they are on balance.

This cycle of the ecosystem dare not be ignored. We may well be reducing the oxygen-producing capacity of the plant world by felling our forests, by paving over the land, by destroying the tropical jungles in an attempt to find minerals or to develop agriculture, and by poisoning the algae in our oceans through DDT.

The Nitrogen Cycle

Still another cycle that must be considered is the nitrogen cycle. We live at the bottom of an ocean of nitrogen since about 79 percent of the air is nitrogen, and we need nitrogen to form one of the basic chemical substances in living protoplasm, protein. Yet we cannot change nitrogen into nitrogen compounds; we must depend on other organisms to do that.

The first step is the conversion of free nitrogen into nitrates. This is accomplished by electrical discharges in the atmosphere (lightning) and also by the nitrogen-fixing bacteria and some of the blue-green algae. Some of the nitrogen-fixing bacteria form nodules on the roots of legumes: this accounts for the fact that many legumes act as "green manure," increasing the nitrate content of the soil. Some of the nitrogen-fixing bacteria are also free living: they are not attached to the roots of legumes and carry on their activity freely in the soil. A number of blue-green algae are also responsible for nitrogen fixation in seas, lakes, rivers, and even to a small extent in the soil.

The nitrates are the simplest form of nitrogen compound which can be utilized by most plants so that the activities of the nitrogen-fixing bacteria and the blue-green algae are extremely important. The green plants take these nitrogen compounds and change them through amino acids into plant protein. This plant protein is then broken down by herbivores and changed into animal protein.

Theoretically animals could use amino acids; this is the simplest type of nitrogen compound that animals and man can utilize. However for all practical purposes we are dependent on plant protein since amino acids do not occur naturally, and ultimately in this cycle man is dependent on the nitrogen-fixing bacteria and the blue-green algae. He has indeed been able to synthesize some nitrogen compounds and use these as an artificial fertilizer for his crops, but the nitrates synthesized by the blue-green algae, the nitrogen-fixing bacteria, and lightning are much larger in quantity than those which he produces synthetically; the synthetic nitrogen compounds are only a supplement to the naturally produced compounds. Any material in the soil which is likely to kill off the nitrogen-fixing bacteria and any material in the water which is likely to damage the blue-green algae is potentially a real hazard to man's own welfare; this is one of the hazards of herbicides and pesticides which are accumulating in our soils and streams.

The Decay Cycle

Another important cycle is the decay cycle. Man is ultimately dependent on the plants for his food, but these can use only relatively simple compounds in their synthetic processes. The various chemical compounds which are needed by plants and therefore also by animals are tied up as complex compounds and are therefore useless until these are broken down to simpler compounds. The bacteria of decay are important in doing this, fungi are also important in this activity, and the various insect scavengers play their role in reducing the complex animal compounds to simpler and therefore usable compounds.

The activities of these bacteria of decay and fungi are extremely important: however they are ordinarily able to break down only naturally occurring substances. A major problem with synthetic substances is that most of them are not "biodegradable," that is, they are not broken down by the bacteria and fungi in the soil. Consequently they accumulate in larger and larger quantities and create a major pollution problem. This is why DDT has accumulated and this is why many of our herbicides accumulate. It is for this reason too that solid wastes tend to accumulate. Iron rusts, and metal cans, while they are unsightly, will eventually disintegrate. Aluminum cans, too, disintegrate though they last much longer than the typical "tin can." Glass bottles are a major problem since glass is a synthetic substance and is not acted on by any of the disintegrating agents in the environment; it is the most durable solid contaminant of the environment.

Inhibiting Decay

One danger of introducing toxic substances into the soil is the damage they may do to the bacteria and fungi of decay. When these attack our foodstuffs we consider them undesirable, and we attempt to destroy them and to interfere with their activities. But by and large their activities are good, and we cannot afford to sterilize our soils, for then we shall soon be buried by the accumulation of plant and animal remains which cannot be broken down and are not consequently made available to other living things which need the chemical elements which they contain.

Man's Arrogance and Lack of Understanding

What else is responsible for environmental crisis? There is no doubt that part of the problem is due to a ruthless and unthinking exploitation of the environment. Man has adopted the attitude that he is above nature, that he can control nature and force it to do his will, and that he can undo any of the damage which he does—once he recognizes it. And so he has been led to exploit his environment, polluting air, water, and soil. He has taken the attitude that the solution to pollution is dilution and has poured solid particles into the air expecting the air currents to dilute them so that they are no longer harmful to him. When he has damaged some areas, as he may by the smelters which destroy the vegetation of surrounding agricultural areas, he has been willing to compensate the owners of these properties for the damage done to their land—money talks and even rights wrongs. The streams and rivers he has regarded as open sewers, depending on nature to undo some of the damage which he does, and being confident that effluents which nature cannot clean up his advanced technology will. At times he has arbitrarily determined that certain streams should serve as open sewers and has not hesitated to pour the waste of millions of people into these sewers; this is the case with the Hudson River in New York. Since it is very apparent even to the unthinking that such water cannot be cleaned up he has arbitrarily reserved other streams and bodies of water for his drinking water. Consequently New York City makes no attempt to take drinking water from the Hudson River which flows along its western border but instead brings relatively unpolluted water through a series of aqueducts down to the city from the Catskill reservoirs.

The "Tragedy of the Commons"

A major factor in our pollution problems has been what Dr. Garrett Hardin has called the "tragedy of the commons." [3] He compares air, water, and soil to the New England commons. Originally this was the common pasturage for the community. In that day all members of the community were permitted to pasture a fixed number of cattle, sheep, and other domestic animals on the commons; each family in the community had its allotment. So long as this allotment was not exceeded the community prospered, and all was well. But there was always the temptation for some individual to exceed his allotment. He reasoned that the profit from the extra sheep or cow which he pastured on the commons would be his own, while any damage to the commons because its carrying capacity was exceeded was distributed among all members of the community. Thus he had the full profit from the extra cow or sheep that he pastured on the commons, whereas the damage done and the loss which he experienced was only a fraction of his gain. For that reason there was always the temptation for exploitation of the commons by some unscrupulous and selfish individual.

Today's "Commons"

The air, the water, and the soil are our commons today. Each of us must breathe the air, must drink the water, and must use the soil for the production of his food. The cost of preventing damage to the commons is substantial. It costs a great deal of money to eliminate smoke pollution, it costs a great deal of money to reduce the pollutants discharged by the internal combustion engines of our automobiles, it costs money to provide primary, secondary, and tertiary treatment for the waste waters from our homes and factories. Many people reason that the cost of cleaning up the water and the air by the individual or the corporation far exceeds the benefits which he will receive, and many people are tempted to reason that they will be benefited financially by forcing the community to assume the costs of correcting the damage which they do to the commons.

Population Problems

Another cause for environmental deterioration in the last third of the 20th century is an increase in our populations and an increase in our standard of living. It is estimated that the world's population is increasing now by about two people every second, 7,000 every hour, 60 to 65 million every year. These figures of population growth are figures of net increases in our population: they represent the birthrate minus death rate.

Each time that you take a breath there are seven more people in the world, seven more people who must be fed, clothed, and sheltered. And it does no good to hold your breath – they keep coming whether or not you breathe. While census figures would indicate that the world's population is a little in excess of 3.5 billion, because of the inadequacy of census figures, a more reasonable estimate of the world's present population is approximately 4 billion.

It is estimated that the world's population at the beginning of the Christian era was somewhere between 200 and 300 million and that at the beginning of the modern era it was about 500 million. In 1950 the world's population is estimated to have been about 2.5 billion.

Not only have populations increased, but the rate of population growth has also increased substantially. Between 1650 and 1750 populations increased at a rate of about 0.3 percent per year but between 1900 and 1950 at a rate of about 0.9 percent. At the present time, the annual rate of population growth is about 1.7 percent. In the more developed countries the rate of population growth is about 1.5 percent but in the underdeveloped about 2.5 percent.

South America has the fastest rate of population growth of all of the continents.

If the present rate of population growth continues we may expect a world population of about 7 billion by the year 2000, and we shall rapidly be reaching the figure of 50 billion, which is the highest responsible estimate of the population-carrying capacity of the globe; this figure will be reached in less than 200 years.

Is it possible these estimates of population are in error? Yes, it is quite likely that they are wrong; it would be very surprising if an extrapolation of this sort would prove to be correct. However, it is more likely that estimates of populations are on the low side rather than being an exaggeration. This has been the history of population predictions. For instance, it was estimated at one time that the population of the United States would stabilize at about 180 million.

The present population of the world is approximately equal to that projected for the year 2000 by Meier, a demographer who gave his estimates of population in 1950.[4] In 1954 the U. N. estimated that the 1980 population of the world would be approximately 4 billion. Today it is clear that this estimate was in error by at least a half billion people. Someone has suggested that this extra half billion people will produce 2,500,000,000 pounds of waste per day or, if this waste had a density equal to that of water, a square mile could be covered to a depth of two feet per day. The heat generated by this extra half billion people would be sufficient to boil an amount of water equal to a lake 10 miles long by 10 miles wide and 10 feet deep every hour of the day.[5]

A few years ago three demographers wrote an article which they titled "Doomsday: Friday 13

November A. D. 2026." Using a slide rule and the highest United Nations predictions of population growth, they calculated that on this day human life would cease on earth because human beings would be squeezed to death.[6]

The "Haves" and the "Have Nots"

One of the critical aspects of the population explosion is the fact that the gap between the "haves" and the "have nots" is widening. The per capita income in the United States in 1960 was $2,300 per year, but in India it was only $70 per year, so the average American is about 32 times as wealthy as the average Indian. Thus if the average American deposited his annual income in a bank and lived off the interest, he would still be better off than the average Indian. The world average for the wealthier nations in 1960 was $1,700 per person per year; for the poorer countries $110 per person per year, so the relative gap averaged 15:1. The gross national product in both the wealthy and poor nations seems to be increasing at a rate of 4 percent per year, but the population increase of 2.5 percent in the poor countries and 1.5 percent in the wealthier countries brings about a continual widening of the gap between them.

The Effects of Poverty

The effects of poverty in shortening of life and in suffering are apparent not only in the poor countries but also in the United States. One study showed that infants of the poor mothers were 15 percent smaller than infants of the nonpoor mothers. Infants from poor families also had multiple anatomical evidences of prenatal undernutrition.

Perinatal mortality rates are higher in the United States than in many other nations; an excess of infants of low birth weight accounts for much of this high perinatal mortality. Low birth weight and perinatal death are much more common in families of low socioeconomic status than in families that are better off. It is believed that undernutrition is the major cause of low birth weight in the group of infants born to poor urban mothers.[7]

It is very likely that this study reflects conditions among the poor in other countries of the world.

An Increase Possible in Food Production

Can we hope to improve the amount of food which we produce? Past experience would indicate that we can. Agricultural production in the United States has been increasing at a rate of about 2 percent a year since 1940 while acreage devoted to food production has actually declined. In the 30 years between 1930 and 1960 the average corn production per acre doubled from 26 to 56 bushels per year. In 1945 we produced 2.5 billion bushels of corn on 77 million acres; in 1965 we produced more than 4 billion bushels on nearly a third fewer acres. In 1945 the average milk yield per cow was 5,000 pounds per year; today it is 8,000 pounds. The number of cows in our dairy herds is 10 million fewer than were needed to supply a smaller market 20 years ago.[8]

In 1934, at the bottom of the depression, U. S. production of commercial broilers was just under 100 million pounds live weight; in 1965 more than seven billion pounds were produced, a 70-fold increase. The live price in 1934 was about 19 cents per pound; in 1965 it was 15 cents.[9]

A century ago seven million farm workers served a total U. S. population of 31 million. By 1910 farms employed 13 million in a nation of 106 million. But today, with a total population of over 200 million, fewer people are employed on farms than in 1910.[10]

There is hope of matching some of this progress on a worldwide scale. Researchers have suggested a number of ways of increasing the supply of protein-rich foods and improving the protein value of grains. Through research in genetics an effort is being made at present to improve the amino acid composition in corn, wheat, and grain sorghums. Two Purdue scientists recently developed a new variety of corn having approximately twice the lysine content of other varieties. The Rockefeller Foundation has undertaken a program to breed a high lysine wheat.[11]

Another approach to improving the biological value of protein is fortification of cereal grains with amino acids produced by chemical synthesis or by fermentation of such carbohydrates as sugar or corn syrup. A fermentation process is now being used commercially in Japan. Chemical synthesis, though not yet important, may take the lead in the future.[12]

Using Plant Proteins

Another approach to improving the diet of people in the developing nations is to have them use protein from seeds, principally soy beans, cotton seeds, and peanuts. Meal produced from these seeds after the oil has been extracted usually has a protein content of about 50 percent as compared to 11 or 12 percent in winter wheat or 8 to 9 percent in corn. Seed oil meal is now used mostly as livestock feed or fertilizer. Each year India produces well over half a million tons of cottonseed meal and 2 million tons of peanut meal, but India's millions of undernourished children are not benefiting from this potential source of protein-rich food.[13]

These suggestions for improving the diet depend on the use of plant materials. The eating of fresh meat even for those who can afford it is a notoriously inefficient way for human beings to get the calories and protein they need; this is evident from a consideration of man's place in the ecosystem and the energy loss that results as we pass from one trophic level in the ecosystem to another. It has been suggested that we may produce synthetic meats by giving plant products the flavor we associate with meat; on a small scale this is already being done.

Farming the Oceans

Some have suggested that we may be able to meet some of the world's needs for food by farming the oceans, the chief source of protein for most nations of the world. Yet this is probably not a likely source of additional food. It is generally believed that the potential fish yield is no more than 100 million tons. The total fish landings for 1967 were just over 60 million tons, and this figure has been increasing at the average rate of about 8 percent per year for the past 25 years.[14] It is clear that while the yield can be still further increased, the resource is not vast; studies of food chains demonstrate this. The open sea—90 percent of the ocean and nearly three fourths of the earth's surface—is essentially a biological desert. It produces a negligible fraction of the world's fish catch at present and has little or no potential for yielding more in the future.

Upwelling regions totaling no more than one tenth of one percent of the ocean's surface, an area roughly the size of California, produce about one half of the world's fish supply. The other half is produced in coastal waters and a few offshore regions of comparably high fertility[15] but subject to pollution.

While a review of developments in increasing food production shows that there is some hope of increasing the available amount of food, it certainly does not suggest we ignore the population problem. We need more research in food production, and we need funds to develop the processes which seem promising. However, such a campaign of increasing the world's food supplies ought be coupled with a program of population limitation. We have had remarkable success with our program of death control: modern medicine has extended the life span not only of Western man but also of people in the underdeveloped countries. But we have seen death control without any comparable birth control.

Changing Population Goals

For many years the goal of population planners was to slow down the rate of population growth. Their slogan was to give men and women the opportunity to have only the number of children they wished. There are now many who believe that we need to set a goal not of a small increase in population but of zero population increase. They point out that even a small population increase, such as we had in medieval times when the annual rate of increase was about 0.3 percent, nevertheless does result in a marked increase in the total number of people. They believe that we must work for a goal of no increase whatsoever, and they believe that if we couple such a program with a program of increasing our food resources we will increase markedly the standard of living of our populations.

Population Quality

One aspect of the problem that needs to be considered is population quality. It is a fact that at the present time the "have not" nations are outbreeding the "haves." It is hard to prove that this is an undesirable situation, and yet it is hardly reasonable to believe that it is a desirable one. As a minimum we ought to seek a situation in which the reproductive rate of both the haves and the have-nots is approximately the same—ideally a rate of zero population increase.

The aspect of population quality cannot be ignored. We have every reason to believe that intelligence is largely inherited, and while everyone will grant that the present measures of intelligence leave much to be desired because of inadequacy of the measuring instruments, it is unlikely that anyone would argue that there is no correlation or a negative correlation between I. Q. tests and intelligence, or that there is a negative correlation or no correlation between intelligence and potential for service to society.

Increasing Populations and Our Environment

We must recognize that our increase in populations has aggravated problems of environmental deterioration. We have been carried away by the myth of progress and have ignored the fact of "spaceship earth." We have managed to develop the idea that growth is good and that a steady-state condition is bad. We are not satisfied unless we can project a 4 percent or 5 percent increase in the gross national product into the foreseeable future. A business must grow and increase its profits, or its management will face serious problems with its stockholders. Yet growth for growth's sake is the ideology of a cancer cell.

The Myth of Progress and Growth

This concern with growth is an inevitable by-product of the Darwinian notion of progress. The philosophy behind Darwinism has suggested that growth and progress are both inevitable and good. Yet this growth inevitably increases our problems.

One example is that of the growing demands for electric power. To satisfy public demand the electric power industry has had to increase its installed capacity at a rate of 6 percent or 7 percent per year for many years, and typical projections suggest a similar rate of increase for the future. No one would suggest the scrapping of our present electric power plants, nor would anyone suggest a moratorium on building new plants provided these do not damage the environment, for we are all deeply indebted to the industry for what it has brought us in comfort, convenience, and leisure.

Yet this projected expansion of our electric power industry and its growth is creating problems and tensions and will do so for the foreseeable future. There are problems involved in site selection: power plants need to be located where they can produce electricity inexpensively, and sometimes these are sites of great natural beauty. There are problems of air pollution, for even with the best equipment it is difficult to prevent all discharge into the atmosphere. More recently we have become aware of the problem of thermal pollution. Power plants, especially the nuclear power plants, use tremendous quantities of water for cooling, and this heated water when returned to a lake or stream changes the entire natural balance. We need to be aware of the effect on environment quality that such aspects of growth are having.

Increased Resource Consumption

Another factor that dare not be neglected is the increasing demands on our resources which our affluent society makes. It is not only that there are more of us as our population grows, but we are actually using the earth's resources more profligately. In 1966 the United States with only 6 percent of the world's population consumed 34 percent of the world's energy production, 29 percent of all steel production, and 17 percent of all timber cut.[16] In the United States electrical energy production during 1965 was 5,000 kilowatt hours per capita. In this same year energy production in Chile was about 680 kilowatt hours per capita, approximately 13 percent that of the United States. Ecuador, one of the more underdeveloped nations in South America, produced less than 100 kilowatt hours of energy per capita, approximately 1.9 percent that of the United States. India, a much larger nation than Ecuador and surely much more highly industrialized, produced even less energy—approximately 50 kilowatt hours per capita, about 1 percent of that of the United States.[17]

Also to be considered is the fact that with our industrialization and with the increase in our standard of living more and more people are moving to the cities. This is true not only in the United

States but throughout the world. At the present time about 38 percent of the world's population lives in urban areas, over 20 percent in cities of 100,000 or more, and about 10 percent in cities of over a million. If the present trend continues, approximately 50 percent of the world's population will be living in urban areas in about 16 years, and extrapolating the present growth rate still farther the total world population will be living in urban areas within 55 years,[18] something which of course will never happen, since the rate of urbanization will one day slow down and halt.

One result of increasing urbanization is crowding, stress, and marked environmental deterioration. This coupled with our enhanced standard of living and with the trend toward a throw-away economy aggravates many of our problems including that of solid wastes. Here in the United States we spread 48 billion rustproof cans and 26 billion nondegradable bottles over the landscape every year. We produce 800 million pounds of trash a day —a great deal of which ends up in our fields, in our parks, and in our forests. Only one third of the billion pounds of paper we use each year is reclaimed. Nine million cars, trucks, and buses are abandoned each year and while many of them are used for scrap a large though undetermined number are left to disintegrate slowly in backyards, in fields, woods, and on the sides of highways.[19]

What needs to be recognized is that we live on "spaceship earth." Our resources are limited even though they are indeed greater than we once recognized them as being. A little more than a hundred years ago the petroleum industry was born in western Pennsylvania. Since that time approximately 71 billion barrels of oil and 214 trillion cubic feet of gas have been produced in this country. The record has been compiled in the face of predictions which date back as far as 1916 that we were rapidly approaching the limit of our resources. Fortunately such predictions have proved unfounded, and the discovery of new petroleum reserves has consistently outpaced demand, with the result that year after year the figures on proven reserves have increased. In addition to its reserves of liquid hydrocarbons the United States has tremendous deposits of oil shale. These deposits, according to the U. S. Department of the Interior, could eventually yield at least a trillion more barrels of oil.[20]

We need to recognize the size of our resources, and we need to be very careful about predicting their depletion because such predictions may be self-defeating if wrong—as they have been in the past. The general public tends to discount *all* predictions of the future when some prove to be in error.

And yet we do live on spaceship earth. While we may not be able to delimit the extent of our reserves it is obvious that they are limited and that we cannot go on using them up. There is certainly nothing wrong with using these resources, but they must not be misused, abused, and wasted.

We need to recognize that we are a part of the ecosystem and need to act accordingly. We have alluded to this earlier. We need to acknowledge the truth of Sir Francis Bacon's aphorism that we cannot command nature except by obeying her. Often, as we shall see, the Christian church is faulted for suggesting that man stands apart from nature. We want to point out later how much this represents a misunderstanding of Biblical statements. Yet there are many who suggest that man is apart from nature. The evolutionary philosophy of Pierre Teilhard de Chardin suggests that man is apart from nature and can live in this way.[21]

Also the philosophy of science's omnipotence and omniscience contributes to this problem. "Scientifically we have only learned to know nature as matter made up of atoms and electrons and not as cosmic processes. Our concrete perceptions of the formative living principles inherent in nature and man are almost completely repressed. Urbanized man under the spell of abstract monetary and physical concepts and locked up in cities is now progressively estranged from a tangible living experience and an awareness of biological principles. The ancient bonds between man and nature are getting weaker. We cannot effectively relate to nature our sense of responsibility, our weakened feelings or principles of ethics when nature is primarily dealt with as systems of electrons, abstract energy, and molecules.... We need to improve our logic and see DNA as a principle of life instead of life as a function of DNA. More than a century ago Goethe rebelled against this one-sided bias of 'mechanistic' science. He understood that form, shape, and pattern—the expressed form in nature and man—were neglected elements in the science of his time. Form as a reality is distinctive from matter. Above all it is expressive —it carries aesthetic and moral content—the meaning of phenomena. Adolf Portmann and other European biologists are clearly developing a combined aesthetic-scientific approach to animal form. We need to emulate and extend such methods in American biology and nature study."[22]

Urbanization and Man's Estrangement

As the quotation cited above indicates, our increasing urbanization as well as the myth of scientific omnipotence has contributed to man's idea that he is separate and apart from nature. A. J. Sharp says much the same thing "Men in primitive groups seem to understand the relationships between themselves and plants better than men in more advanced cultures. There seems to be an almost direct correlation between the degree of the civilization of a society and its failure to appreciate this dependence. This lack of awareness is further enhanced when a civilized society changes from an agricultural to an industrial economy. Industrialization seems effectively to isolate the individual from the primary plant resources on which his culture is based."[23] Kendeigh says "the laws of nature apply to man as they do to animals. There are no exceptions. If he can come to understand what they are and how they work, he will know better what to anticipate concerning their effects on himself. He cannot ignore the dynamic forces of the environment with impunity, but being blessed with an intelligence far above that of other animals he can guard against them or alleviate their effects to his own advantage."[24]

Social Illnesses

Another effect of man's failure to recognize his place as a part of the ecosystem is the change from concern with physical diseases to social illnesses. McGill says man is radically changing his environment. His ills are mostly social. "He often flees to quiet retreats by the lake or the mountainside.

He speaks of 'getting away from it all,' he dreams of a small farm, he goes to analysts, he fills up ever-expanding mental hospitals. He suffers from 'nerves,' 'stress,' and frustrations he cannot articulate. . . . Neither comfort, mink stoles, automobiles with power of hundreds of horses, membership in key clubs, diamonds, luxurious air conditioning, automatic heating, television, nor pleasures unlimited serve to console or reassure him. . . . His free will enables him to change his environment but he cannot change his genes, and therefore his mind has to adapt easily to what he has done to himself. Sitting in an automated palace he envies his grandfather or an earlier ancestor who lived in a log cabin or small farmhouse and subsisted off the land. God seems farther and farther away from him." [25]

Because man has regarded himself as above nature he has often done things without counting the ultimate cost. "Much of the thrift of yesterday turns out today to be prodigality because the price tag did not include all of the social and economic costs . . . although competent to run a primitive world man may not be competent to 'manage the more complicated and closely integrated world which he is, for the first time, powerful enough to destroy.' "[26]

Good Intentions Are Not Enough

Consider just one example of how man may inadvertently have done damage in not considering the complexity of the ecosystem, and in his desire to manage nature has actually defeated his own ends. Conservationists have long been interested in maintaining the giant sequoias of the West, yet it is now evident that the longtime fire prevention and suppression policies have created two conditions in many of the national park sequoia groves both of which are contrary to established park policy. Fuels once consumed by regular fires have accumulated to unprecedented amounts, which present uncommonly great fire hazards. Plant succession which was previously reversed by fire has progressed in most groves to the point that the giant sequoia is almost completely unsuccessful in reseeding itself. Where protection from fire is complete there is a tendency for the sequoia to disappear.[27] Thus man with the best of intentions had harmed rather than protected, because he has not understood the complexities of the ecosystem.

NOTES TO CHAPTER 2

1. Berton Roueché, *Eleven Blue Men* (Boston: Little, Brown, 1947), pp. 194-215.
2. James P. Dixon and James P. Lodge, "Air Conservation Report Reflects National Concern," *Science*, 148 (1965): 1066.
3. Garrett Hardin, "The Tragedy of the Commons," *Science*, 162 (1968): 1243-8.
4. Richard Rosenbloom, "Book Review," *Science*, 154 (1966): 252.
5. H. E. Hoelscher, "Technology and Social Change," *Science*, 166 (1969): 69.
6. Heinz von Foerster, Patricia M. Mora, and Lawrence Amiot, "Doomsday: Friday 13 November, A. D. 2026," *Science*, 132 (1960): 1291-5.
7. R. L. Naeye, M. M. Diener, and W. S. Dellinger, "Urban Poverty: Effects on Prenatal Nutrition," *Science*, 166 (1969): 1026.
8. Orville L. Freeman, "Agriculture/2000—Science in the Service of Man," *The Science Teacher*, 34 (5): 49.
9. T. C. Byerly, "Efficiency of Feed Conversion," *Science*, 157 (1967): 890.
10. Freeman, p. 48.
11. Luther J. Carter, "World's Food Supply: Problems and Prospects," *Science*, 155 (1967): 57.
12. Ibid.

13. Ibid.
14. John H. Ryther, "Photosynthesis and Fish Production in the Sea," *Science,* 166 (1969): 76.
15. Ibid., p. 75.
16. Luther J. Carter, "The Population Crisis: Rising Concern at Home," *Science,* 166 (1969): 725.
17. Hoelscher, 69.
18. Walter G. Peter, III, "Sex Education: A Key to the Population Crisis," *BioScience,* 20 (1970): 173.
19. Carter, "Population . . ." etc., loc. cit.
20. John Bivins, "Petroleum Reserves," *Science,* 139 (1963): 798.
21. Frederick Elder, *Crisis in Eden* (Nashville: Abingdon, 1970), pp. 62-8.
22. *Nature Study* 21 (2): 7 (Summer, 1967).
23. A. J. Sharp, "The Compleat Botanist," *Science,* 146 (1964): 746.
24. S. Charles Kendeigh, "The Ecology of Man, the Animal," *BioScience,* 15 (1965): 523.
25. Ralph McGill, "Editorial," *BioScience,* 15 (1965): 762.
26. "Public Policy and Environmental Administration," *BioScience,* 17 (1967): 884.
27. Richard J. Hartesveldt, H. Thomas Harvey, Howard S. Shellhammer, and Ronald E. Stecker, "Sequoias' Dependence on Fire," *Science,* 166 (1969): 552.

Chapter 3

Upsetting the Balance of Nature

Another cause for our ecological crisis today is the upset which man has so often brought about in the balance of nature. The Christian viewing the complexity of these balances cannot help but be impressed with God's wisdom in setting them up. They are so intricate that they could not have developed by chance.

Of course we must recognize that the balance of nature which God established at creation is a dynamic balance. It is not exact but oscillates within a given range. In every natural situation there are imbalances which nature finds it possible to correct. However, in many cases man has introduced exotic plants and animals which have upset the balance of nature by inflicting changes greater than nature could cope with. In some cases these introductions have been deliberate with the idea that man would be able to improve on the balance of nature. In other cases they have been inadvertent.

The Rabbit in Australia

When the early settlers came to Australia they found there no placental mammals except the dingo or wild dog and a few species of rodents. Coming from Europe as they did they remembered the fine hunting provided by the rabbit there. And so in an attempt to improve on nature Thomas Austin imported some 24 European rabbits in 1859. The result was unfortunate, for there were no natural enemies to keep the rabbits in check. They multiplied beyond all expectation and did serious damage, destroying the grass upon which the sheep fed.

At first an attempt was made to control them by building a rabbit-proof fence across the continent in Queensland, but this proved useless, for either the rabbits were already through it or it was not truly rabbit proof. Then an attempt was made to reduce their numbers by a system of bounties, but again this proved unsuccessful. Only in recent years has a solution been found, and this is the introduction of a virus disease, myxomatosis, which kills the rabbits and keeps their numbers in check. Even this may not be the final answer, for after a short time the rabbits developed a virus-resistant strain which allowed them to begin multiplying again. Then the virus developed a mutant which was able to attack the virus-resistant strains of rabbits so that at the present time an uneasy new balance is being struck.

The present reduction in rabbit numbers has had great benefits. Grasslands once ravaged by erosion and hills grazed to the soil for decades are now miraculously clothed with green. During the 1952–53 season, one of the first when the myxomatosis campaign was carried out, the sheep industry alone showed an increased productivity worth about $84 million.

It is interesting that the introduction of the rabbit virus into Europe had its effects there. A French doctor, concerned about the damage the rabbits were doing to his shrubbery, imported a culture of the virus and injected it into a few rabbits that he had trapped. These were then released. The result has been a reduction in the rabbit population in France, in the neighboring countries on the continent, and in England. What the total effect will be is still being debated. There has been a loss of meat supply which was formerly available to the common people and upon which they depended. Others report that the loss of meat has been more than offset by the increase in garden crops.

The Macquarie Islands

Rabbits were introduced in the Macquarie Islands to improve the food resources there. Soon they began to destroy the crops. To repair this damage cats were released. For a time this was successful as the cats preyed on the rabbits, but once the rabbits had been disposed of, the cats attacked the seabirds whose eggs the natives greatly prized. Once again man intervened. This time he released dogs to keep the cats in check. But the dogs preferred seals which were important in supplementing the natives' food supply. And so at present attempts are being made to destroy the dogs that man introduced to destroy the cats that man introduced to destroy the rabbits that man introduced.

Coyotes in Colorado

Nor have we in the United States been immune from such practices. The ranchers of the Toponas district of Colorado, wishing to save their sheep and cattle, carried out a campaign to exterminate the coyotes that were attacking their lambs and young calves. The coyotes disappeared but the ranchers noticed that their pasture land was no longer able to support as many animals as before. The reason was that, with nothing to stop them, rabbits, gophers, and other rodents began to attack the meadows. Now the Colorado ranchers are encouraging coyotes to breed.

Carp

In the 70s of the last century the German carp was introduced into the streams and lakes of the Midwest. This fish is a valuable food fish in Europe and when first introduced carp sold for as high as $80 a pair. Their introduction was actually promoted by the Bureau of Fisheries. The streams in which they lived in Europe are clear, but in the United States they found streams and lakes with muddy bottoms. In feeding on the vegetation of their new homes the carp swallowed muck with their food, and their flesh took on an unpleasant muddy taste. In the spring great schools of carp come up the streams to spawn, rooting up the bottoms. The result is that much of the vegetation is destroyed. Because of the destruction of vegetation and because the water is now so muddy, most of the native game fish have disappeared from the streams in which the carp are abundant.[1]

The Sparrow Campaign in Red China

Man has still not learned, and today we still face problems brought about by upsetting the balance of nature. In Red China a campaign was recently carried on to exterminate the European tree sparrow, a member of the weaver bird family. A notoriously ravenous feeder on cereal crops this sparrow was designated as one of the "four evils" and marked for immediate extermination in Red China.

However, in 1960 a campaign was begun for the protection and conservation of the sparrows which the government had vowed to exterminate. The Chinese Communists had not realized how complex the balance of nature is, for the almost complete extermination of the sparrow was soon followed by a serious outbreak of cereal insects.[2]

Czechoslovakia

At the close of World War II the economic planners decided that a large aluminum processing plant was to be located somewhere in Slovakia. Military authorities strongly preferred a site in one of the narrow valleys of the lower or middle Tatra. Some scientists demurred on the ground that smelter fumes would cause damage, but they had no hard facts to present, and the plant was constructed in the valley. At the time the available fuel was a local coal high in sulphur content. Within a few months it became evident that the valley had a typical atmospheric inversion layer that trapped fumes in the valley. As the sulphur dioxide content of the air increased all plants were killed, and all animals that were not killed were driven away. The mill workers had to live at a considerable distance, and at times they even had to wear gas masks in order to continue to work.[3]

The Aswan Dam

Another recent problem is that raised by the construction of the Aswan high dam on the Nile. It is generally agreed that this dam will have a number of effects, most of which will be unfavorable. The effect on the sardine fisheries of the eastern Mediterranean will be catastrophic, since the nitrates and phosphates which the Nile River now brings down regularly into the Mediterranean will no longer be there, and we shall have to expect a marked decline in the sardine take. Already with the dam only partially in operation, the catch has been reduced by 50 percent. Another effect will no doubt be the rapid spread through the Egyptian population of infestation with parasitic blood flukes or schistosomes whose intermediate hosts are snails. The snails spread through the irrigation canals, and these will be extended in order to utilize the water which the dam makes available.[4] Finally the very brief life of reservoirs impounding irrigation waters must be considered. When a river carries a heavy burden of silt such as the Nile does, the reservoir fills quickly with mud and silt, and the question arises what will happen when the reservoir is full of these and will no longer hold an appreciable amount of water.

Thus the increase in the Egyptian per capita income from $130 to $142 a year seems a high price to pay for the damage that is being done by this structure—especially in the light of the fact that it probably will not serve for a very long time.[5]

The Soviet Union

Soviet plans to divert southward three giant Siberian rivers now flowing into the Arctic Ocean could trigger worldwide climatic changes according to studies which have been conducted by British scientists. Mediterranean type weather could move farther north in Europe while desert areas could grow in central Asia. It could even mean greater deserts in the United States although this is less likely.

Work has already gotten under way on a 15-year scheme to redirect the waters of the Pechora, the Ob, and the Yenisei Rivers toward the desert region around the Caspian and Aral Seas. This plan would deprive the Arctic of about half the fresh water now flowing into it, and it is believed would shift the world's climatic belts farther north. Mediterranean areas might experience North African conditions and Mediterranean type weather would move farther north in Europe. The program would have great advantages to the Russians: they would be able to irrigate about five million acres of arid land and would be able to drain 150 million acres of salt marshes. Yet these rivers are important for maintaining the ice cover of the Arctic Ocean and supply much of the water that keeps the top layer of the ocean comparatively fresh so that it freezes more easily. If the supply were reduced or cut off there could be large scale melting of the Arctic waters.[6]

The Dust Bowl of the Midwest

The dust storms of the thirties are another example of man's upsetting the balance of nature. The tough sod which held the soil was destroyed by overgrazing and by plowing for food crops with the result that topsoil blew away during the regular periods of drought. Yet we continue to follow practices which have the same effect, unwittingly turning grasslands and food producing areas into wasteland and desert. The Sahara Desert for example is still increasing by 40,000 acres a year.[7]

The Gypsy Moth

An example of an accidental introduction of an import was that of the gypsy moth which reached the United States in 1886. It was hoped that by using this moth the native silk industry could be established. It escaped accidentally, and the moth has proved to be a serious pest. It feeds especially on native shade trees. In 1953 1,500,000 acres of trees in New England were defoliated by the pest. As late as 1957 $5 million was spent by state and federal governments in an attempt to control it. At the present time some control is being gained by the use of DDT and other pesticides.

Lake Michigan

Another area which has suffered through upsetting the balance of nature is that of our freshwater Great Lakes. We shall refer later (Chapter 5) to the eutrophication (increase in nutrients favoring plant over animal life) which is going on in Lake Erie and Lake Michigan, where overnourishing by man's nutrients and industrial wastes results in oxygen deficiency. Another aspect is the change in the species composition of the fish. These changes have been taking place since records were first kept, but they have been particularly abrupt in the last 10 to 20 years. Some changes are conspicuously related to changes in the environment such as eutrophication, but others have been brought about or very much influenced by fishing practices and by the introduction of exotic species.

In Lake Michigan, for example, the lake trout has practically disappeared as a climax predator. This has resulted not only from overfishing but through the accidental introduction of the lamprey. As a result of the removal of the lake trout two species of deep water ciscoes have become extinct. Four others are near extinction and the remaining species of those which were once the forage base for the lake trout are undergoing rapid biological changes which some fish biologists interpret as indicating that a complete population collapse may be imminent.[8]

The niche occupied by these species has been taken over by the alewife which was first reported in Lake Michigan in 1949 and became established in the lake about 10 years later. Now it dominates the entire lake much to the regret of not only fishermen but of those who live on the shores of the lake because these fish are characterized by repeated die-offs fouling the beaches. Thus in less than 20 years an entirely new lake population without a climax predator exists.

Some steps are being taken to restore the original conditions. The sea lamprey which provided the primary impetus for the upset by its extreme depredation of larger fish has almost been brought under control by methods which proved successful in Lake Superior. The lake trout is being reestablished, and the exotic coho salmon is being introduced as a climax predator to establish a new population balance. But would it not have been far better to avoid upsetting the balance of nature to begin with?

The whole sea lamprey story is an interesting one. It was possible for this fish parasite to reach the Great Lakes because of the construction of the Welland Canal which enabled it to invade the western Great Lakes. It took approximately 100 years for the population explosion of the sea lamprey to occur, but when it did it resulted in the decimation of the white fish and lake trout fisheries. In the last 12 years the United States and Canada have contributed approximately $16 million for control measures and research on this problem. This figure does not include the millions lost by the fishing industry of the region or the contribution of the various state conservation and fishery research programs.[9]

Snails in Florida

The state of Florida is presently involved in an intensive control program to destroy thousands of fist-sized snails, *Achatina fulica,* found in a 13-square-block area in Miami. The origin of the snails is unknown, but the Department of Agriculture believes that several of the snails were brought from Hawaii 3 years ago by a child as a gift for his grandmother. The snails multiply hermaphroditically—each individual has both male and female sex organs—and reach adulthood in about 30 days. A cornmeal bait containing calcium arsenate and metaldehyde is being spread by the Florida State Department of Agriculture to kill the snails. How successful this will be is unknown at present.[10]

Domestic Animals

Domestic animals cause many problems resulting in upsetting natural balances. They destroy a great deal of vegetation and compete with the native animals, usually to the extinction of the native species over their range. Particularly dangerous has been the introduction of goats onto various islands. Mariners have often introduced these hoping thereby to provide themselves with a ready source of fresh meat. The usual effect has been to upset the balance of nature on these islands and destroy much of the native vegetation, with the consequent starvation of native animals.

Much of the superb and valuable indigenous growth of Hawaii was destroyed by livestock which Cook and Vancouver introduced before 1900. In order to protect the soils and water supply a quick forest of eucalypts and other foreign trees has been introduced, but the dry, undecomposed litter produced by these does little to promote the infiltration and storage of vital ground water in the islands.[11] Thus the balance of nature has been upset by the introduction of domestic animals and wrong vegetation.

New Mexico is having a problem with gerbils. The gerbil is a small burrowing rodent similar in many ways to a hamster. They are sometimes sold as pets though their sale in the state of New Mexico is forbidden. Their native habitat is similar to many areas of New Mexico. If they are released there they may well survive and reproduce in numbers that have made them pests elsewhere. A swarm of gerbils in India totally destroyed all crops over an area of about 8,000 square miles in 1878, literally mowing down the standing stalks of grain.

The Crown of Thorns Starfish

Probably the most recent example of upsetting the balance of nature is the plague of crown of thorns starfish – really a sea star – which developed recently in the South Pacific. These starfish grow to be two feet across and have an insatiable appetite for coral. They have 16 arms instead of the normal five found in starfish, and each of them can destroy a square yard of coral a night. More than 50 percent of the living coral at Guam has been destroyed. The Great Barrier Reef north and east of Australia has become a favorite target, and some 140 miles there have been destroyed. Seventy percent of Tinian's reef has been wiped out. The spines of this starfish are venomous; this is a hazard to human beings.

Reports of infestation by this animal began about 1963 when the phenomenon was first observed on a major scale in the Great Barrier Reef off Australia. The exact cause of the outbreak is not known. It has been suggested that the population explosion may have resulted from a decrease in the number of conchs which feed on the crown of thorn starfish. But this explanation is not generally accepted today. It is also possible that man has destroyed some of the filter feeders through poisoning the waters with DDT. These filter feeders may have been responsible for eating the larvae of the crown of thorn starfish. Another theory is that dredging and blasting have destroyed the filter feeders or has provided a good place in which the larvae of these seastars could develop. In any case it is agreed that upsets by man are responsible for this plague.[12]

A New Panama Canal

In recent years considerable concern has been expressed over the possible hazards of a sea level canal across the Isthmus of Panama connecting the Atlantic and the Pacific. The present Panama Canal, being essentially fresh water, does not permit the movement of marine species between the two oceans to any great extent, but it is feared that a sea level canal may upset the balance of nature in either of the two oceans by introducing exotic species which will thrive and destroy native species.[13]

Problems in Brazil

There have been still other recent developments which point to the hazards of upsetting the balance of nature. In Brazil's Amazon basin an agricultural colony was established in the late 1940s at the confluence of the Madeira and the Madrede Dios Rivers. Rain forests were cut away and the land was tilled with disastrous consequences. Because the soils of the region are rich in iron and aluminum oxides they are highly subject to a process called laterization which can quickly turn cleared lands into rocky brushy barrens. The colonists' efforts to survive were unbelievable. They cultivated their fields among blocks of laterite and found the soils they were working had compacted into rock in 5 years time.

The Kariba Dam

The Kariba Dam constructed in the 1950s on the Zambezi River between Zambia and Rhodesia has formed a huge 1,700-square-mile reservoir rivaled in size only by the reservoirs created by the Volta Dam in Ghana and the Aswan Dam in the United Arab Republic. Planned largely to generate electrical power, the Kariba project led to a variety of unanticipated problems. Serious soil erosion is occurring because farmers who were flooded off their former lands have refused to change their traditional farming methods to suit conditions in relocation areas. Also the vegetation along the shores of Kariba Lake has created the type of habitat attractive to the tsetse fly—aggravating the pest control problem in this area where the population is heavily dependent on animal protein. Moreover, aquatic plants invaded Kariba Lake and within 4 years of the dam's completion covered more than one tenth of the lake surface, interfering with commercial fishing and creating other difficulties.[14]

Man and Natural Balances

There is general agreement that we need to go slow in introducing exotic organisms into the environment, and we probably ought to extend this concept to any synthetic material, a subject we shall discuss later. The balance of nature which God has established is a good one; man tampers with it at great risk. This is not to suggest that man dare never interfere with a natural situation; it is obvious that our agricultural practices represent an alteration of natural conditions. If man is to exercise his rule over nature, he must make some alterations. However, he needs to be sure that the alterations he makes will benefit him and not prove harmful in the long run. Probably as a matter of principle we need to prove that something is harmless before it is introduced into the environment. Our usual procedure has been to permit the introduction of any substance until it was proven to be harmful. We need to reverse this procedure and insist that its harmlessness be demonstrated before it is introduced.

We also need to recognize the dynamic quality of the balance of nature. As we pointed out there is a range within which nature is able to operate. Nature does correct many of the imbalances which are introduced by natural forces and also many of those which are introduced by man. Our concern today with the mounting environmental crisis is that we may be adding the straw that breaks the camel's back. We had an illustration of the possibilities of total breakdown in the electric power blackout of 1965 which blacked out most of the East Coast. The relatively minor failure of a single switch triggered this catastrophe. Similarly a simple upset in the natural world could lead to disastrous consequences.

NOTES TO CHAPTER 3

1. John W. Klotz, *Modern Science in the Christian Life,* (St. Louis: Concordia, 1961), pp. 126-32.
2. Tien-Hsi Cheng, "Insect Control in Mainland China," *Science,* 140 (1963): 273 f.
3. Bentley Glass, "For Full Technological Assessment," *Science,* 165 (1969): 755.
4. Luther J. Carter, "Development in the Poor Nations: How to Avoid Fouling the Nest," *Science,* 163 (1969): 1047.
5. Dael Wolfle, "The Use of Arid Lands," *Science,* 164 (1969): 1347.
6. *Fort Wayne Journal Gazette,* 108 (Feb. 24, 1970): 2.
7. Wolfle, loc. cit.
8. Grayce A. Finger, "Overexploited Animal Populations," *Science,* 154 (1966): 544 f.
9. Ira Rubinoff, "Central American Sea-Level Canal: Possible Biological Effects," *Science,* 161 (1968): 858.
10. *Science,* 166 (1969): 355.
11. Paul B. Sears, "A Pacific Tropical Botanical Garden," *Science,* 147 (1965): 241.
12. "Crown of Thorns Investigation," *BioScience,* 20 (1970): 113.
13. John C. Briggs, "Panama's Sea Level Canal," *Science,* 162 (1968): 511.
14. Carter, loc. cit.

Chapter 4

Air Pollution

In considering pollution problems we need to understand just what pollutants are. It is clear that they are the residues of things we make use of and throw away. Our whole economy is based on taking natural resources, converting them into things that are consumer products, selling them to the consumer, and then forgetting about them. But there are no consumers—only users. The user employs the product, sometimes changes it in form, but does not consume it. He just discards it. This practice of discarding things creates the residues that pollute, at an increasing cost to the consumer and to his community. The problem is aggravated not only by our increased use of materials but by our increasing populations. As the earth becomes more crowded there is no longer an "away," and one person's trashbasket becomes another person's living space.

Air Pollution

Probably the first type of pollution of which men became aware was that of air pollution. Seneca described it already A. D. 61, and our modern environmental crisis began with 62 deaths reported in Belgium's Meuse Valley in 1930, 22 deaths in Donora, Pennsylvania, in 1948, and 4,000 deaths in London in 1952. All parts of the world are affected by air pollution; the rain in some parts of Sweden is about as acid as Coca Cola with a pH of 2.5. This acidity is attributed to the oxides of sulphur that belch out of England's and northern Europe's industries and are carried to Sweden.

Studies indicate that the twelve worst American cities in air pollution are New York, Chicago, Philadelphia, Los Angeles and Long Beach, Cleveland, Pittsburgh, Boston, Newark, Detroit, St. Louis, Gary, Hammond, East Chicago, and Akron.

We are finally recognizing how extensive the damage to our atmosphere is. In 1960, some 23,300,000 tons of sulphur dioxide were released in the air. Most of this waste came from the combustion of coal. It is believed that by 1980 the figure for sulphur dioxide may reach 36,000,000 tons. In Los Angeles alone the daily discharge of sulphur dioxide is 455 tons. Most of this comes from home heating plants, from factories, and from electrical generating plants. To burn this fossil fuel 3,000 cubic miles of air are used each year.[1]

In Pittsburgh some 610 tons of dust settle each year per square mile, 1 1/2 tons per day in summer, 2 1/4 tons per day in winter. This dust is made up of soot or carbon particles, iron oxide, which constitutes about 20 percent of the dust, and silicon dioxide, which constitutes about 16 percent. In addition 14 or more other metal oxides have been found. This does not include the oxides of sulphur and phosphorus, which are gases.[2]

Someone has said it is with the coming of man that a vast hole seems to open in nature, a vast black whirlpool spinning faster and faster, consuming flesh, stone, soil, minerals, sucking down the lightning, wrenching power from the atom until the ancient sounds of nature are drowned in the cacophony of something which is no longer nature, something instead which is loose and knocking at the world's heart, something demonic and no longer planned—escaped it may be—spewed out of nature, contending in a final giant's game against its master.

Increases in Carbon Dioxide

The burning of fossil fuels—coal, oil, and natural gas stored from the past—has raised the carbon dioxide content of the air over 10 percent since the end of the last century. Each year the clean atmosphere increases its carbon dioxide content by 0.25 percent, according to records at Mauna Loa, Hawaii, and Antarctica. This growth rate is only half the estimated quantity from burning of fuels; the remaining half of the output is absorbed mainly into the oceans.[3]

More carbon dioxide is not necessarily detrimental; trees and other plants can theoretically grow faster in the presence of higher concentrations. However, the increased carbon dioxide content may result in increasing temperatures through the "greenhouse" effect. Carbon dioxide acts as a blanket to reduce the amount of heat which is radiated at night, and this may have the effect of increasing our temperatures. Counterbalancing this is the fact that increased pollution may result in a reduction in insolation, the amount of heat energy which reaches the earth from the sun. It may well be that these two are very close to being in balance, the one effecting a temperature increase, the other a temperature decrease.

Particle Numbers

One problem of air pollution is the small size of some of the particles which are emitted. When in good operating condition the effluent from the auto exhaust pipe is quite invisible. However, if one measures the number of particles emitted by an idling automobile engine, it is on the order of 1 hundred billion particles per second. Air pollution control may actually result in an increase in the number of such invisible particles, since air pollution laws are usually aimed at the control of visible smoke fumes; the laws in most communities are designed to force industrial plants to install electrical precipitators, scrubbers, and other smoke control devices. However, in some instances it is possible to pass the effluent from an industrial process through a hot flame, an afterburner, to vaporize it. Thus as with the automobile the pollution plume becomes invisible. The concentration of tiny particles is so high, however, that a conglomeration or clumping often occurs and the knowledgeable observer will be able to detect the plume downwind from the offending source.

Under such conditions the use of an afterburner may be undesirable, since in addition to making the particles much smaller than they would normally be, they now have a longer residence time in the atmosphere because of their smaller size. Moreover, an afterburner will generate nitrogen oxide, a poisonous gas which serves as a catalyst for particle growth involving unburned gasoline vapor.[4]

Smoking and Air Pollution

In the process of smoking the individual draws into his lungs at least 10 million smoke particles per cubic centimeter. This is a concentration somewhere between 10 and 100 times greater than is encountered in a badly polluted urban area such as Los Angeles or New York City.[5]

Carbon monoxide is another potential pollutant. One hundred and twenty parts per million of carbon monoxide for an hour causes inactivation of 5 percent of the body's hemoglobin and commonly leads to dizziness, headache, and lassitude.

In New York City alone automobile traffic produces 8.3 million pounds of carbon monoxide each day. Each automobile emits about 1/6 pound of carbon monoxide per mile of travel at 25 miles per hour and about 1/3 pound per mile of travel at 10 miles an hour. According to the best estimate, 20 million pounds of carbon monoxide per day were emitted by motor vehicles in Los Angeles in 1967.[6]

It is estimated that there are 42,000 parts per million of carbon monoxide in cigarette smoke. Fortunately the smoker rarely inhales this much. However in a poorly ventilated, smoke-filled room concentrations can reach several hundred parts per million.[7]

Cigarette smoke is a problem in other respects as well as in its carbon monoxide. Levels of five parts per million of nitrogen dioxide (NO) are considered dangerous: cigarette smoke has 250 parts per million. Ten parts per million of hydrogen cyanide is dangerous; the concentration in cigarette smoke is 1,600 parts per million. The toxic effect of cigarette smoke may be enhanced by other environmental factors: among 283 asbestos workers who smoked, 24 of the 78 deaths which occurred were due to lung cancer; yet among the 87 non-smokers who died none died of lung cancer.[8]

Pollutant Quantities

Each year it is estimated that we are pouring at least 183 million tons of toxic matter into the air. Such figures are meaningless to most people, but it represents 2/3 ton for every man, woman, and child in America, and the prospect for the future looks even blacker. By 1980 we will have a third more people in our cities. We will have 40 percent more automobiles and trucks, and we will be burning half again as much fuel.

Weather Modification

Pollution is believed to have brought about some modification in weather. Over the past several years a number of unusual snow and rain storms have been observed in the east central part of New York State. These storms consist of extremely small precipitation particles. When in the form of snow, the particles are like snow dust, having cross sections ranging from 0.02 cm to 0.05 cm. In the form of droplets they are often even smaller in diameter, at times being so tiny that they drift rather than fall toward the earth. When collected on clean plastic sheets, the precipitation is found to consist of badly polluted water.

Another case in point seems to exist around LaPorte, Indiana, about 30 miles downwind from the smoky mills of Gary and South Chicago. From 1951 to 1965 LaPorte had 31 percent more precipitation and 38 percent more days with hail than nearby communities. Year to year correlations were found between the rainfall in LaPorte and variations in steel production and hazy days in Chicago.[9]

Health Hazards

There is clear evidence of health hazards brought about by air pollution. We referred earlier to the smog deaths and illnesses in Donora, Pennsylvania; New York; London; and the Meuse Valley. Emphysema is the fastest growing cause of death today. In the 10-year period 1950–1959 death among males from emphysema rose from 1.5 per 100,000 to 8 per 100,000. In 1962 more than 12,000 persons died of emphysema in the United States, and this total has been increasing steadily. At the same time it is estimated that each month a thousand more workers are forced to retire prematurely onto the social security rolls because of the disease.[10]

On Staten Island it has been found that residents of the northern half of the island have a much higher cancer death rate than individuals who live on the southern half of the island. The northern half of Staten Island is affected by pollutants from the Bayonne-Elizabeth industrial complex. Lung cancer kills men over 45 years of age at a rate of 55 per 100,000 on the northern half of the island but at a rate of only 40 per 100,000 on the southern half of the island. Many of the men who live on the island commute to New York City, but most of the women spend all of the average day on the island. With women the cancer death rate is twice as high on the northern half of the island as on the southern half.

Studies conducted in Buffalo, New York, showed

a definite correlation between high levels of air pollution and mortality rates. Researchers found the incidence of stomach cancer three times greater in a highly polluted area than in one with low pollution levels. Links between chronic bronchitis and emphysema on one hand and air pollution on the other are becoming clearer.[11] Sudden increases in respiratory illness and death were dramatically evident in the London smog of 1952 when an excess mortality of 4,000 persons was recorded in the 5-week period. Subsequent investigation has shown that changes in the atmospheric concentration of oxidants, carbon monoxide, sulphur dioxide, and oxides of nitrogen are significantly related to hospital admission rates and length of stay for respiratory and circulatory conditions.

Symptoms of chronic pulmonary disease have been found more frequently in areas of greater air pollution. Anatomic emphysema increases in prevalence above the age of 40 and has been shown to be more common in the sample from a heavily industrialized urban community with higher pollution than in one from a purely agricultural city with lower pollution. The "cause," like that of other chronic conditions, appears to be an interaction of several environmental and host factors including age, climate, and smoking as well as air pollution.

So far as bronchitis and emphysema are concerned there were 18,763 more deaths attributed to these two causes in the United States in 1966 than 10 years earlier, an increase of almost 2 1/2 times and one observed in all ages above thirty-five.[12] If only one half of the increased deaths could be attributed to air pollution this would still amount to nearly 1,000 additional deaths each year. Chronic bronchitis and emphysema are typical chronic diseases in which years of increased disability usually precede death.

Possible Genetic Damage

Robert Shapiro, associate professor of chemistry at New York University, has found in laboratory experiments that sodium bisulfite, the form which sulphur dioxide takes when absorbed into the body, damages nucleic acids in a number of ways which have "serious biological implications." Dr. Shapiro reports that sodium bisulfite latches on to two of the components of nucleic acid making it non-functional. When the bisulfite molecule breaks away from the nucleic acid it converts cytosine into uracil. These are two of the "letters" of the "genetic code." It is necessary to change only one out of millions of units in a nucleic acid in order to cause a mutation; sulphur dioxide converts as much as 90 percent of the cytosines into uracils in yeast RNA under physiological conditions. Sulphur dioxide is the first chemical found that can specifically change one natural nucleic acid component into another.[13]

Damage to Materials

Air pollutants create other problems as well. They abrade, corrode, tarnish, soil, erode, crack, weaken, and discolor materials of all varieties. Sulphur pollution attacks and destroys even the most durable materials. Steel corrodes two to four times faster in urban and industrial areas than it does in rural areas where less sulphur-bearing coal and oil are burned. Cleopatra's needle has deteriorated more since its arrival in New York in 1881 than it did during the 3,000 years it spent in Egypt. It is estimated that in New York a family of four spends $800 a year as an invisible tax to undo damage and clean away the dirt left by air pollution.

Crop Damage

Another effect of polluted air is damage to and destruction of vegetation. Airborne pollutants in some parts of the nation are doing more damage to plants and crops than even bad weather or the ravages of insects. The price tag for agricultural losses is estimated in the neighborhood of $500,000,000 annually. More than 11,000 square miles of farmland are affected in California already, and the threat is spreading into the rich San Joaquin Valley which contains five of the leading farm counties in America.[14]

In the garden state of New Jersey pollution injury to vegetation has occurred in every single county, and damage has been reported to at least 36 commercial crops. Nearly a score of other states including Florida, Idaho, Montana, Pennsylvania, Oregon, North Carolina, Tennessee, Utah, Washington, and Colorado have suffered serious crop losses due to air pollution.

Smog damage has been found as far as 60 miles from Los Angeles, and one scientist believes that pollution drifting from large cities in the central valley of California is threatening vegetation in such world famous national parks as Sequoia.[15]

Air pollution now places very real restrictions on the types of vegetation that may be raised in many sections of the country. In the metropolitan areas of California, for example, photochemical smog has made it impossible to raise orchids and the once prosperous growers of these regions have had to relocate in remote rural areas.

In Polk County, Florida, a dozen phosphate fertilizer plants which produce 31 percent of the world's output of phosphate pollute the air with tons of fluorides which have ruined thousands of acres of citrus groves and withered large areas of shrubbery. Once Florida's leading cattle producer, the county has suffered serious losses to its herds from fluoride pollutants settling on its pasture lands. Registered steers that once were worth $3,000 a head have had to be sold for as little as $50 or slaughtered as a hopeless loss because the fields in which they once grazed contentedly are now destroyed by pollution.[16]

Meeting Pollution Problems

The increasing industrialization and the growth of our economy are likely to increase our pollution problems unless some drastic action is taken. The health problem cannot be ignored. A specially designed "clean-air room" to treat victims of air pollution has been established at New York City's St. Vincent Hospital. The new facility, one of the first in the country, will provide purified, humidified air and will be equipped with oxygen supplies and respirators. With the anticipated continued increase in the national air pollution picture, Dr. Stephen M. Ayres, director of the hospital's cardio-pulmonary laboratory, believes that all hospitals should set aside clean-air areas.[17]

The demand for clean air has hit the electric power industry very hard. Conventional power plants using fossil fuels pollute the air, and the newly designed nuclear plants, as we shall see, add to problems of thermal pollution. All large coal burning power plants in the United States use the pulverized fuel technique in which powdered coal is blown into furnaces with very large volumes. This method produces much particulate matter and leaves the waste gases in an oxidized state in a large volume of air. The particulate matter can be collected with electrostatic precipitators which utilize well-established technology, but they are usually larger than the furnace and are expensive to install. For instance, the precipitator for the thousand megawatt plant at Ravenswood, New York, cost $10 million.

The search for an economical method of collecting sulphur from pulverized fuel combustion has been going on for 30 years, but there are still no efficient devices for collecting sulphur on any large power plants now in operation. At a recent power generation symposium, Arthur Squires of the City College of New York presented a convincing case for a broader approach to the problem. Noting that power production by coal was expected to be the major source of energy for at least the rest of the century and that sulphur emission is perhaps the major pollutant from coal combustion, he argued that major research efforts on new methods of combustion are entirely justified. As long as the sulphur is spread through a large volume of air and is in the form of sulphur dioxide (conditions inherent in the pulverized fuel combustion method), the search for methods to remove it are likely to fail. However Squires noted that as early as 1942 a method of pulsing air and steam to burn coal had been developed in Germany. In the mid-50s the method was modified in France so that it was applicable to the large furnaces required for modern power production. Squires believes that it would be possible to combine several features of presently used techniques in such a way that the sulphur will be released as hydrogen sulfide in a small volume of air. This would require the development of techniques for extracting the sulphur under pressure, but Squires believes that this difficulty would be outweighed by the advantage of having the sulphur in its reduced form in much smaller volume.

The Air Quality Act

What can be done? We need to recognize the extent of the pollution problem, its economic cost, and the health problems which it generates. The Air Quality Act of 1967 is a big step forward in the control of air pollution. It authorizes more federal funds to combat air pollution than we have spent in all 180 years of our history. The system hinges on the designation of air quality control regions, in which groups of communities either in the same or different states share a common air pollution problem, and on the development and enforcement of air quality standards for such regions. The Department of Health, Education, and Welfare will designate specific air quality control regions. These will be designated on the basis of factors which suggest that a group of communities should be treated as a unit for the purpose of setting and implementing air quality standards. Factors to be considered include meteorological, topographical, social, and political considerations, jurisdictional boundaries, the extent of urban and industrial concentrations, and the nature and location of air pollution sources. The work of designating these regions is now in progress following the preliminary step of defining broad atmospheric areas.

The Department will also develop and publish air pollution criteria indicating the extent to which air pollution is harmful to health and damaging to property and giving detailed information on techniques for the prevention and control of air pollution, including data on comparative costs and effectiveness of their application. As soon as air quality criteria and data on control technology are made available for a pollutant or class of pollutants, states will be expected to begin developing air quality standards and plans for implementation of the standards. This step has already been taken in many places.

If the Secretary of Health, Education, and Welfare finds that the air quality standards and plans for implementation of the standards in an air quality control region are consistent with the provisions of the Air Quality Act, then those standards and plans will take effect. But if a state fails to establish standards or if the secretary finds that the standards are not consistent with the act, he can initiate action to insure that appropriate standards are set.

The Need for Management Agencies

Strict enforcement will be required under this act, but there is another area which should be developed and which may be helpful. We need not only regulatory agencies but also management agencies to serve as resource organizations to which municipalities and industries can turn for assistance. These agencies would not be charged with enforcement but rather with research and with the responsibility for supplying engineering help.

Paying the Costs

It is obvious that we shall have to accept inconvenience and higher costs as the price of cleaner air. We shall no longer be able to afford the luxury of open-air burning of trash and leaves. We shall have to limit the kinds of fuels we use to heat our homes to those that are "smokeless." We shall have to pay more for the things we buy, because the costs industry incurs in reducing and eliminating pollution will have to be added to the selling prices of its products. But we are already paying for pollution in costs of cleaning drapes, curtains, and clothes; in the more rapid deterioration of the things we buy; and in illness and shortened lives. In the end, spending money to stop pollution may really save money, reduce illnesses, and add to our life expectancies.

NOTES TO CHAPTER 4

1. Frederick Sargeant II, "Adaptive Strategy for Air Pollution," *BioScience,* 17 (1967): 691-7.
2. *BioScience,* 17 (1967): 879.
3. Lester Machta, "Winds, Pollution, and the Wilderness," *The Living Wilderness,* 33 (Summer 1969), 4.
4. Vincent J. Schaefer, "Some Effects of Air Pollution on Our Environment," *BioScience,* 19 (1969): 896.
5. Ibid.
6. John R. Goldsmith, and Stephen A. Landaw, "Carbon Monoxide and Human Health," *Science,* 162 (1968): 1352.
7. Philip H. Abelson, "A Damaging Source of Air Pollution," *Science,* 158 (1967): 1527.

8. Ibid.
9. Earl G. Droessler, "Weather Modification," *Science,* 162 (1968): 287.
10. John T. Middleton, *We Can Have Clean Air* (Waukesha, Wis.: Country Beautiful Foundation, 1969), pp. 1 f.
11. *Chicago Tribune,* Feb. 9, 1970, Section 1A, p. 2.
12. Amasa Ford, "Casualties of Our Time," *Science,* 167 (1970): 258.
13. *BioScience,* 20 (1970): 240.
14. Middleton, loc. cit.
15. Ibid., loc. cit.
16. Ibid., loc. cit.
17. *BioScience,* 19 (1969): 930.

Chapter 5

Water Pollution

A second area of major pollution concern is that of water pollution. We became concerned with this problem around the turn of the century when we found people dying of waterborne diseases such as typhoid fever. This led to the development of water purification plants which supply most of our cities with water that is reasonably free from pathogenic organisms. Since water must serve two purposes—supplying water for drinking, industry, irrigation, and agriculture; and removing human wastes and those of our factories—we have usually not been too concerned so long as the wastes could be removed from drinking water or were not obviously harmful to health. Indeed, in highly industrialized areas we arbitrarily decided that some streams should be used as open sewers to remove our wastes and others should be used as sources of drinking water.

The Hudson in the eastern United States and the Rhine in western Europe are two classic examples of badly polluted streams. The Hudson near New York City is little more than an evil smelling sewer whose water is foul not only with sewage but also with oil, orange rinds, detergents, thick wads of pulp, and sundry other human and industrial wastes. Only a few eels manage to survive on the river's bottom. A few years ago three small boys who ate a watermelon which they fished from the river came down with typhoid fever even though we regard this disease as quite unusual.

Upstream, the Albany pool which stretches from that city north to the Troy Dam is even fouler. Between Fort Edward and the Troy Dam alone the Hudson receives the raw wastes from 600,000 people. Local residents know when summer has arrived by the hydrogen sulfide stench which accompanies the thaw of sludge on the river's bottom. The water here is 50 times more polluted than water approved for swimming.

It is hard to convince people living around Albany or New York City that to early explorers the Hudson valley once gave off only "sweet perfume," or that there is still anything worthwhile in the river today. Yet in the stretch of the river south of Albany but north of New York City the Hudson manages to clean itself and teems with life. Schools of striped bass, sturgeon, shad, herring, perch, large mouth bass, giant carp, and various other fishes thrive the year around. Ducks and geese find the river here hospitable enough to visit during their migration. Some towns even use it for drinking water.

The pollution begins again at approximately West Point. Much of the pollution north of Albany is due to pulp wastes: thick grey mats bob on the surface of the water there like shredded mattresses. South of West Point industrial wastes, oil, and raw sewage are freely dumped into the river.

The Fabled Rhine

The Rhine River rises in Switzerland. When it reaches Basle, the first major city on its long journey to the North Sea, it has a pollution content of 10 to 15 units per cubic meter. By the time it reaches Strasbourg in France, nearly 100 miles downstream, the pollution rate is up to 2,000 units, and it is unsafe to swim in the Rhine. Five hundred miles farther downstream the river has passed a dozen cities of more than 300,000 population, and several of up to a million or more. It has also picked up the wastes of the Moselle and Ruhr Rivers. When it reaches Rotterdam and the open sea the Rhine's pollution rating has soared to a colossal 1.5 million units per cubic meter.

Other Polluted Rivers

These are not the only rivers that are badly affected; the Missouri River near St. Joseph, Missouri, is so foul that crows can float downstream on grease balls, hides, or mats of hair being dumped into the river by packing plants.

All told the cities of the United States discharge 65 billion gallons of sewage each year directly into streams and rivers. This does not count the amount of treated sewage; more than half the cities of 2,500 or more do not have a good sewage disposal plant. While this figure is somewhat deceptive since the larger cities have sewage disposal plants and most of the cities included in the figure above are only slightly larger than 2,500 in population, the problem is still acute. Most of the present sewage disposal plants were devised for conditions which existed 40 years ago, when many streams could handle a small amount of pollution, and for populations which existed at that time when our cities were much smaller. It is estimated that even with efficient waste treatment our effluents will be sufficient by 1980 to consume all the oxygen in all the dry weather flow of the 22 river basins in the United States.

Even the effluent from a good sewage disposal plant requires river water for dilution. It is generally said that for every gallon of effluent there should be seven gallons of river water. Yet in one of our midwestern states it was reported that in one dry period there were seven gallons of effluent for every gallon of river water and that a short distance downstream this effluent and river water were met by the effluent from another sewage disposal plant at the rate of three gallons of effluent for every gallon of mixed river water and effluent from upstream.

Primary, Secondary, and Tertiary Sewage Treatment

It is generally agreed that as a minimum a sewage disposal plant must provide secondary sewage treatment or its equivalent. Primary treatment of sewage takes out sticks and rags and other debris plus about 45 to 50 percent of the settleable solids and 30 to 40 percent of the oxygen-consuming sewage wastes. Chlorination to kill bacteria but not viruses follows. Even though primary treatment is inadequate, 30 percent of the U. S. municipalities stop here; according to one authority this is little better than throwing slops out of the second-story window.

Secondary sewage treatment uses bacteria to remove up to 90 percent of the organic matter, and the discharged water appears clean. However, the process does not remove the nutrients that cause eutrophication or the many chemicals. In fact the

10 percent or more of the wastes which escape treatment even in the finest and best operated plants can heavily pollute a lake or stream.

The ultimate goal in controlling water pollution is tertiary treatment which removes up to 99 percent of the pollutants remaining after primary and secondary treatment.

Storm Sewers and Sanitary Sewers

Complicating the problem of treating sewage is the fact that in many cities the storm sewers and sanitary sewers are combined. The municipal sewage disposal plant is generally equipped to handle the flow from the sanitary sewers or even to handle this plus the flow from a moderate rain storm. However, when there is a downpour the combined flow from the sanitary and storm sewers is more than the plant can handle. In these cases the usual practice is to open the gates and dump the untreated sewage—diluted indeed by the heavy rainfall—into the nearest stream.

The sewage disposal problem is further complicated by the fact that new industries are creating exotic wastes which travel downstream unchanged for hundreds of miles. These new and unusual contaminents include DDT, *o*-nitrochlorobenzene, pyridine, detergents, diphenyl ether, kerosene, nitriles, and a variety of substituted benzenes. These substances are not removed by waste treatment or water purification processes. Moreover, because they are synthetic and do not occur in nature, they are not broken down by the bacteria and fungi in the soil. We do not know what all the effects of these pollutants are when they appear in drinking water.

Detergents

One example of newer substances creating problems is the one which detergents created not too long ago. The majority of the older types of detergents leave ABS, alkyl benzene sulfonate, as a residue in the water. Water begins to foam when ABS occurs at the rate of one part per million, and public health standards permit up to one half part per million in water that is considered fit for human consumption.

Unfortunately sewage treatment plants do not remove ABS from the water, since it is a synthetic substance. The detergents which came to the market in the 1930s and which proved more effective than soap in cleaning power clogged sewer systems and streams with foam.

This problem was solved in 1965, when the detergent makers, urged on by angry consumers and congressmen, switched the chemical composition from ABS to linear alkylate sulfonate, which was more easily broken down by bacteria. Bio-degradable detergents are now the only sort one can buy. However, these continue to contribute to the eutrophication problems to which we shall refer shortly, since they are high in phosphates.

A more direct threat to the safety of consumers today may be created by the newest detergent innovation, enzymes. Enzyme products are being investigated at the present time by the Federal Trade Commission, the Food and Drug Administration, and the Public Health Service. The enzymes used are the products of *Bacillus subtilis* which produces them as a part of the digestive process. Theoretically the enzymes attack protein stains which other detergents are not able to remove, and they are advertised as being particularly effective against blood, urine, meat juices, chocolate, and so on.

The flour-like dust of these enzymes in the air in manufacturing plants eats away at the skin of the workers' hands and sensitizes their lungs, producing an allergic reaction with hay fever or asthma-like symptoms. These are considered an industrial hazard by government agencies and by the detergent makers, but there is no such evidence at present on enzyme hazards to housewives, because the detergents and presoaks contain only small amounts of the substance, anywhere from 0.3 percent to 1 percent. Just what the result of these investigations will be no one is in a position to know at present.

Eutrophication

Detergents have been major contributors to the eutrophication which has gone on in many of our streams and lakes. Perhaps the best known example of this is what has happened in Lake Erie, which many regard at present as a dead lake. Lake Erie is the victim of the introduction of excessive nutrients which accelerate the natural aging of such a body of water. The excessive enrichment or eutrophication is brought on not only by the phosphates introduced by detergents but also by human, animal, and industrial wastes which are introduced into the water.

Eutrophication results first in an overwhelmingly excessive growth of algae and other aquatic plants. It might at first seem that this is a desirable situation, since ultimately fish and other aquatic animals are dependent upon the food which the algae and aquatic plants supply.

However this growth of algae chokes the open waters and makes the water nonpotable. Subsequently the algae decompose, foul the air, and consume the deep water oxygen so vital for fish and animal life. In eutrophication fish families change from trout to warm water bass and perch to plant-eating types and finally to bottom feeders. The balance of a lake or body of water is ultimately upset, because the bacteria are unable to convert the dead organic matter into plant and animal food. The balance is further upset in our northern or temperate zones, because bacteria grow only during the summer, and sewage and garbage are dumped into our lakes and streams all year round.

The major problem in Lake Erie seems to be the problem of phosphates which have been dumped in by sewage treatment plants. Phosphates, which are a major ingredient of detergents—also of the new types of bio-degradable detergents—act as fertilizer for the type of algae which are growing in the lake and which are drawing off oxygen.

Removing these phosphates requires tertiary treatment of sewage; this is the reason for the increasing demands that sewage plants provide this type of treatment. Unfortunately this is a very expensive process.

It is estimated that almost 5 billion pounds of detergents are used every year in the United States.

Carl L. Klein, Assistant Secretary of the Interior for Water Quality and Research, says that even without the phosphates in detergents, 680 million pounds of phosphates from other sources would continue to contaminate the water. He believes that some inland waters may already contain so much phosphate that control of future inputs will not retard eutrophication. The Department of the Interior seeks to deal with all of the phosphates in municipal wastes regardless of their origin. Efforts have therefore been concentrated on the development and demonstration of phosphate removal technology at municipal sewage disposal plants.

While Lake Erie is close to being a dead lake, its situation may be less serious than that of Lake Michigan, even though Lake Michigan is not as highly contaminated as Lake Erie. It is believed that the natural flow of water through Lake Erie and Lake Ontario will be sufficient to remove 90 percent of the wastes present in 20 years if pollution is stopped at once. This is because Lake Erie is a relatively shallow lake with a flow-to-volume ratio of 0.38 per year. Because of this high flow-to-volume ratio the average retention time of its water is only 2.6 years. Lake Michigan, which is rapidly going the way of Lake Erie, has an average retention time of 30.8 years, and it is for this reason that the situation there is much more serious: it is believed that hundreds of years would be required to displace the pollution in Lake Michigan.

Another problem with Lake Michigan is that it is not a flow-through lake. Much of the water remains there almost indefinitely. Some scientists believe that this is favorable; they suggest that the deep waters do not readily mix with the coastal waters because of the differences in currents caused by the earth's rotation, and they feel this may forestall the threatened eutrophication of Lake Michigan. Such disagreement is an example of some of the complexities of the problem and a reason why greater research is necessary. It is also a cogent reason for stopping pollution before it gets out of hand.

Lake Superior has an average water retention time of 189 years so that pollution problems there are potentially even more serious. Fortunately as yet Lakes Superior and Huron show little change due to man's activity except for the accumulation of insecticides in fish and bird life and taconite tailings.

Lake Superior

Lake Superior, one of the cleanest fresh water lakes in the world, is threatened by the disposal of 59,000 tons of finely pulverized taconite tailings into the lake at Silver Bay, Minnesota. This material appears in a colloidal form in Lake Superior and finds its way down into the other Great Lakes. Not only does it increase the colloidal material in the lakes, but it is also important in distributing trace metals.

Part of the problem is that the conversion of low grade rock into usable iron ore is an expensive operation and must be competitive. The particular problem seems to be not so much the colloidal nature of these tailings but that they introduce trace elements into the lake water. These are known to be essential for the growth of algae. The low temperatures of Lake Superior seem to retard the growth of algae in the lake except when the water warms up; plankton blooms have been found only in late August and early September though the green water containing the colloidal particles has been found at other periods of the year. For that reason these taconite accumulations are not too serious a threat at present to Lake Superior, but they are a threat to the other lakes as they reach them.

The well known 1967 spring mass mortality of fish, mostly alewives, in southern Lake Michigan appears to be related to a bloom of toxic blue-green algae. Water masses in Lake Michigan remained cold longer than usual into late spring that year and then suddenly warmed up. Unusually strong stormy winds in the late spring caused the water to be stirred up and the bottom water to be overturned. With the turnover polluted sediments from the bottom were brought into suspension and solution and may have provided enough enrichment to support the bloom of blue-green algae, which gave rise to toxins in the algae and in suspension and in solution in the water.

Thirty-eight people aboard an ore boat from Duluth traveling southward in Lake Michigan at the beginning of the fish kill drank water pumped aboard from the contaminated water which had killed the fish. Although the water had been chlorinated, 35 of the 38 people became very ill. There was no indication of food poisoning but their symptoms were those characteristic of blue-green algae toxicity.[1]

Animal Wastes

Another aspect of water pollution is the problem of animal wastes. These are left in the fields or spread on the fields to act as fertilizers. They are a particular problem in the spring because they accumulate and remain in a frozen condition over the winter and are washed into the nearest stream by the spring freshets.

The development of cattle feed lots, where steers are fed intensively in an operation comparable to that of our broiler factories, is likely to aggravate the situation. The number of cattle raised in feed lots will increase proportionately in the years ahead, since this seems to be a more efficient way of raising them. However this will require special attention to the disposal of the animal waste produced.

Sediment

A further aspect of water pollution is the accumulation of sediment. Appreciable amounts of suspended matter are reported in oceans, but at the present time these are restricted to a few areas off the Atlantic and Pacific coasts. Particles that escape estuaries or are discharged by rivers into the shelf region tend to travel shoreward rather than seaward, so that at present there does not appear to be too great a threat to the open oceans.

Sediment is a major effect of soil erosion, but it is also a factor in water pollution since it interferes markedly with fish life. Its presence close to the coast is serious because this is a major breeding ground for fish.

Fish Kills

Water pollution as a whole does great damage to our fish population. In 1968 the Department of the Interior's Federal Water Pollution Control Administration estimated that 15,236,000 fish perished as a result of water pollution, an increase of 31 percent over the 1967 fatalities. While it is likely that different reporting practices, variation in weather, and other factors could partially be responsible for the increase, the report does indicate that our lakes, rivers, and streams are being poisoned by many highly toxic and dangerous substances. Fish deaths attributed to pollution in 1968 were the third highest reported since the annual census began in 1960, being surpassed only by the 15,910,000 fatality figure reported in 1961 and the 18,387,000 fish reported killed in 1964.[2]

The largest single fish kill in 1968 occurred in the Allegheny River at Bruin, Pennsylvania, where 4,029,000 fish were wiped out after a petroleum refinery company's lagoon overflowed into another pond, breaking down its walls and releasing chemicals into the water. An inadequate sewage plant in Mobile, Alabama, caused the year's second largest kill: effluent from the overloaded treatment plant lowered the oxygen content in a 2-mile stretch of the Dog River, suffocating a little over a million fish.

Animal Kills

Organisms other than fish also suffer from water pollution. In a recent year the Division of Wildlife of the Ohio Department of Natural Resources investigated 58 pollution cases which killed more than 794,900 animals. The greatest animal kills came from metal manufacturing sources, many of which use cyanide, with other manufacturing sources and coal mining second and third. Over 58,000 wild animals were killed in Ohio from sewage pollution alone.

Aquatic Weeds

Another problem of water pollution is that of a rapid growth of aquatic weeds. These obstruct water flow, increase evaporation, cause large losses of water through transpiration, and prevent proper drainage of the land. Weeds may interfere with navigation, prevent fishing and recreation, depress real estate values, and present health hazards. In the western United States it is reported that 17 states lose 1,966,000 acre feet of irrigation water annually because of aquatic and ditch bank weeds. This water, valued conservatively at $20 an acre foot, is worth over $39,000,000.

Like so many instances of upsetting the balance of nature, many of these weed problems are the result of the introduction of exotics. One of the worst of the aquatic weeds is the water hyacinth, a native of South America. These plants increase rapidly by vegetative reproduction. Water hyacinth was first reported in the Congo River in 1952 and in less than three years it had spread 960 miles from Leopoldville to Stanleyville. In 1954 it had already begun to block transportation. By 1957 over one million dollars had been spent in trying to clean up the river, but in spite of this, it was reported that water hyacinths were still floating past Leopoldville on the way to the sea at the rate of 136 metric tons per hour. Water hyacinth consumes and wastes tremendous quantities of water through the leaves.

A similar problem was created by the water fern on Lake Kariba on the Zambezi River. The dam was closed late in 1958, and in May 1959 floating mats of the water fern were reported in the center of the lake. The mats grew in size and moved about with the wind. Before the end of the year a significant portion of the water near the shore was covered. By 1960, less than two years after the dam was closed, it was estimated that the mass of water fern covered 10 percent of the surface area of the lake.[3]

Oil Slicks

Another major problem is that of oil slicks in our oceans. Oil pollution has been a human problem for most of the century, but it took the grounding of the tanker *Torrey Canyon* and the blowout of the well off the coast of Santa Barbara to draw public attention to the major problems that can arise in the production and shipping of petroleum. A fivefold increase in oil production is expected by 1980, and the potential for large-scale pollution can increase faster because of changes in location and shipping practices.

Crude oil is not a single chemical but a collection of hundreds of substances of widely differing properties and toxicities. Paul Galtsoff of the Bureau of Commercial Fisheries at Woods Hole, Massachusetts, stated recently that oil in sea water should be regarded not as an ordinary pollutant but as a dynamic system actively reacting with the environment. It is still impossible to predict the behavior of specific oil spills, and little is known about the long-term effects of oil in a marine environment. Most of the current research is directed toward the immediate problem of handling oil spills and preventing their spread.

In 1966, some 700,000,000 tons of oil, about half of the world's total ocean tonnage was shipped in 3,281 tankers. It is difficult to estimate how much of this oil is spilled, but an estimate from the Woods Hole Oceanographic Institute estimates that somewhere between a million and a hundred million tons of oil are added to our oceans each year. The major sources of this spilled oil are handling errors, leaks from natural deposits, tanker and barge accidents, and illegal tanker bilge washings. Normal techniques of transferring oil to small coastal tankers, barges, and shore facilities are a chronic source of coastal oil. The Massachusetts Division of Natural Resources says that in Boston Harbor alone a spill involving several tons of oil can be expected every third week.

Less frequent but more spectacular are leaks from offshore deposits. These can occur naturally, but they have been associated with drilling operations since the 1930s, when fields in the Gulf of Mexico were opened. Tanker accidents are similar to well blowouts in that an occasional major catastrophe highlights a constant source of contamination. The grounding of the *Torrey Canyon* off the southwest coast of England in March 1967 was simply the most dramatic example of a tanker accident that on a worldwide basis occurs more than once a week. Finally deliberate dumping of bilge washings adds a considerable but unknown amount of oil to the oceanic environment.

It is believed that pumping and shipping operations will continue to expand for the next few decades. Large deposits have been discovered in Alaska and Canada, and test studies indicate the probable presence of oil in the North Sea, the Persian Gulf, and Indonesia. Oil from these sources will be transported through pipelines and by gigantic tankers. The SS *Manhattan* was specially strengthened for travel in the ice and fought her way through the Canadian Arctic in September 1969. The test was generally deemed successful, and plans are being made for a fleet of six 250,000-ton vessels which are to be built for year-round service.

The *Torrey Canyon,* considered a large ship at the time of her grounding, had a displacement of 127,000 tons.

After coming in contact with water, crude oil spreads rapidly into a thin layer and the lighter fractions evaporate. In protected areas the oil often becomes adsorbed on particulate matter and sinks, but in open seas it tends to remain on the surface where wind and wave action aid in further evaporation. Some oil dissolves in sea water and some is oxidized, but the hundreds of species of bacteria, yeasts, and molds that attack different fractions of hydrocarbons under a variety of physical conditions are primarily responsible for oil degradation. Bacteria found in open seas tend to degrade only straight chain hydrocarbons of moderate molecular weights so that branched chain hydrocarbons of high molecular weight form chunks which may persist for a long time.

Damage to Seabirds

The most visible victims of ocean oil are seabirds. It is impossible even to guess how many are killed each year, and about the only thing known for sure is that once oiled, very few birds survive. After the *Torrey Canyon* disaster 5,711 birds were cleaned off, but only 150 of them returned to health and were released. Of these it is believed that about 37 died within one month after release. Very few of the 1,500 diving birds cleaned after the Santa Barbara blowout survived. Most deaths are believed to be the result of disease such as pneumonia which attacks the birds after they are weakened by the physical effects of the oil. The high death rate of cleaned birds suggests long-term toxicity of oil.

The other effects of oil spills are unknown. Some studies indicate that visible damage to organisms other than birds is relatively slight. However, one study, conducted on a small, previously unpolluted cove on the Pacific coast of Baja California, Mexico, indicates that the ecology of an area such as this may be radically changed. In March 1957 a tanker grounded at the mouth of this cove and blocked most of its entrance. All signs of oil disappeared sometime between November 1957 and May 1958. A few species returned within two months, but two years elapsed before significant improvements were noted. Four years after the accident sea urchin and abalone populations were still greatly reduced, and 10 years later several species had still not returned.

Sometimes efforts to eliminate pollution are worse than the pollution itself; this was the case in the *Torrey Canyon* disaster when the aromatic hydrocarbons that were used to dissolve detergents in an effort to disperse the oil spill caused much of the damage.[4]

Every effort will have to be made in the future to reduce the amount of oil spills. However, so long as we are dependent on oil we shall have to expect them. We need research on the full effects of these spills, and we need to develop techniques to confine them.

Estuaries

Another water pollution problem is the contamination of our estuaries. These are important because they are the breeding grounds of so many of our fish and marine organisms. A good example of the problems of estuaries is the situation in Galveston Bay. Extending over 533 square miles, it is the largest of the estuaries on the Texas coast. The bay still supports major commercial and sport fisheries, and oysters, shrimp, and crabs; redfish, sea trout, and other finfish are plentiful. However nearly half of the bay is now closed to oyster harvesting because of pollution, although fortunately the most productive oyster reefs are still open to harvesting. Since the bay is an important nursery for shrimp and certain fish such as croakers, anchovies, and menhaden, which spend part of their life cycle in the Gulf of Mexico, conditions that damage the bay will hurt fishing in the Gulf.

Galveston Bay, like other estuaries, is a brackish body of water which in general becomes less salty toward the head of the bay where there is an important inflow of fresh water from tributaries. For oysters brackish water is essential, as they can survive in neither fresh water nor sea water. The juvenile forms of shrimp and finfish such as menhaden also require brackish water, although these species do best in the lower salinities found near the head of the bay. The bay waters are enriched by nutrients brought in by the tributaries which are flushed out of the shallows and marshes by tidal action, and this too helps to account for the abundance of marine life which this estuary supports.

The bay environment has experienced and is still undergoing radical changes. Shell dredgers have removed most of the shell from the bay, often taking exposed shell as well as shell underlying a heavy layer of silt. Shell is valuable in highway construction and for other uses, and from it fortunes have been made. Little shell dredging is now being done, but until recently the dredgers were taking millions of cubic yards of shell from the bay each year. Not only were shell reefs destroyed, but in some cases the dredging and washing of shell caused the silting up of parts of the reef bearing live oysters.

Livingston Reservoir, which has been built on the Trinity River by the city of Houston and the Trinity River Authority, will store water largely for diversion to Houston. Its effect on the bay will be twofold. First, the flow of fresh water into the bay will be reduced, salinities will increase, and the production of shrimp, oysters, and other marine life may be hurt. Second, while most of the water diverted to Houston will be returned later by way of a ship canal it will be returned in a polluted condition. Still other diversions may be in the offing.

Marshlands are being lost in the bay area. The Wallisville Dam on the Trinity River is intended to open the lower river to navigation and keep salt water from intruding upstream to the intakes of the water supply and irrigation systems. Because the dam has little storage capacity the project will never substantially reduce the flow of fresh water into the bay, but it is eliminating 20,000 acres of brackish ponds, sloughs, marshes, and bottomland, nearly all of which biologists of the U. S. Fish and Wildlife Service regard as prime shrimp and finfish nursery grounds with an annual productive capacity of no less than $300 an acre and probably more.

The Houston ship canal, which follows Buffalo Bayou from Houston to Galveston Bay, ranks as one of the filthiest stretches of water in the United States. Industrial pollutants and huge volumes of poorly treated domestic sewage from Houston and its suburbs are imposing on the channel a daily waste load that is the equivalent of the raw sewage of a city of 2 to 3 million people. Often dissolved oxygen is totally lacking in much of the channel. Wastes from the ship channel are by far the bay's worst pollution problem. In addition to this the city of Galveston discharges one and one-half million gallons of raw sewage into the bay each day.[5]

Water Needs

The whole problem of water pollution is compounded by our increasing need for water. Water consumption today runs between 75 and 100 billion gallons per day; by 1990 it is estimated that our requirements will increase to somewhere between 150 and 200 billion gallons. The average home consumes about 60 gallons per person per day. Industry uses 900 gallons per person per day and agriculture uses 700 gallons per person per day. Some 375 gallons of water are required to produce just one pound of flour. Much of this water is gotten from surface streams, but a great deal is also secured from deep wells which tap the aquifer, the underground water table. Here there is a major difficulty because pumping is removing water faster than the aquifer is recharged. Near Lubbock, Texas, the productive aquifer has been drained. Salt water has intruded into the fresh water aquifers at Baltimore, Maryland; Galveston, Texas; Tampa, Florida; and Long Beach, California. In the suburbs around Chicago deeper and deeper wells have been necessary, and it appears that soon the entire Chicago area will have to get its water from Lake Michigan.

Another problem has been the subsidence of land in Mexico City and in some parts of the central valley of California because of pumping from deep wells. Problems have arisen in areas it would almost seem would have no problems at all. Portland, Oregon, is bisected by the Willamette River and is flanked by the mighty Columbia River; yet Portland has serious water problems apparently brought on by use of large quantities of water for air conditioning.

These underground waters are generally considered to be much purer than surface water; yet it is apparent that if the aquifer is recharged from polluted streams it will itself become polluted. In one instance near Denver, Colorado, chemical wastes were discharged into holding ponds for 12 years, beginning in 1943. It was not until 1951 that any contamination was noted, but in that year some of the water pumped for irrigation was contaminated by the chemical wastes that had been stored, even though this was several miles down gradient from the pond. By 1961 about 30 square miles had been contaminated.

Much of the use of water by industry involves its use in cooling processes, and this water might well be pumped out into the aquifer to recharge it. At present in most cases it is dumped into the nearest surface stream where it does not help the aquifer.

Reusing Water

It is apparent that we shall soon have to get accustomed to reusing water, and there is certainly nothing wrong with this even though it may strike us at first as being an undesirable procedure. It is estimated that already at present the water in the Mississippi is used several times as it journeys from St. Paul to New Orleans. It is taken into a water purification plant, used for drinking purposes and for other home and commercial purposes, then discharged through the community sewage disposal plant into the river. This effluent then is picked up by the downstream water purification plant, and the process is repeated. Since water must be reused, it is essential that it be cleaned up as much as possible before it is dumped into the nearest stream.

The Water Quality Act

In 1965 a Water Quality Act was passed by Congress. This act establishes a framework of federal law and regulation within which each state is to protect the interstate and coastal waters under its jurisdiction. The act provides for the state to adopt water quality standards which are subject to federal review and approval, but the standards are to be enforced by the states themselves unless federal help is requested or unless the state fails to carry out its enforcement responsibilities. In the case of pollution occurring in a single state the Federal Government is to have authority to undertake enforcement action only at the request of the governor of that state or upon finding that shellfish beds are being contaminated.

In adopting water quality standards the states are to classify waters according to the uses made of them; to establish the quality criteria for dissolved oxygen, temperature, acidity, and the like appropriate to the use classification; to fix an "implementation schedule" by which polluters, whether industries or municipalities, must provide for the "best practical treatment" of their effluents not later than 1975; and finally to be ready to enforce their standards.

All 50 states now have water quality standards

that have been approved at least in part by the Secretary of the Interior, but at the beginning of 1970 only 14 states had standards which the federal water pollution control administration had not found deficient in one area or another. Nearly half the states have not yet agreed to protect waters already of high quality against any degree of degradation; this is a major problem since in some states the standards have been accepted as minimum and permission has been given to polluters to reduce high quality water to these minimum standards. Some states have also adopted standards considered too permissive with respect to certain quality criteria.

Following the passage of the Water Quality Act, legislation authorizing appropriations of $3.4 billion over a 4-year period for matching grants for the construction of waste treatment facilities was passed, but actual appropriations for such grants were modest until 1969 when Congress provided $800 million, largely in response to pressure from the grass roots. Over the next 5 years it is believed that somewhere between 30 and 50 billion dollars will be needed to deal with the water pollution problems of the nation.

NOTES TO CHAPTER 5

1. Louis G. Williams, "Should Some Beneficial Uses of Public Waterways be Illegitimate," *BioScience,* 18 (1968): 36 f.
2. *American Biology Teacher,* 32 (1970): 42.
3. L. G. Holm, L. W. Weldon, R. D. Blackburn, "Aquatic Weeds," *Science,* 166 (1969): 700 f.
4. Robert W. Holcomb, "Oil in the Ecosystem," *Science,* 166 (1969): 204 f.
5. Luther J. Carter, "Galveston Bay: Test Case of an Estuary in Crisis," *Science,* 167 (1970): 1102-7.

Chapter 6

Other Types of Pollution

Another problem of environmental deterioration is that of environmental deterioration of the soil. Of all of the early centers of civilization the only one that has not been seriously affected by soil erosion is the Nile Valley. As we have pointed out earlier, Mesopotamia was originally an alluvial plateau between the Tigris and the Euphrates. The source of these two rivers was originally forested, and the rainfall equalled that needed today for the growing of tobacco, cotton, and grapes. Clearing the land disturbed the water cycle and also gave rise to torrential floods in the Tigris and Euphrates valleys.[1] The alluvial soil here was destroyed basically by being forced to support the military conquests of ancient empire builders.

Before 2000 B. C. the Babylonians harvested two crops of grain per year and grazed sheep on the land between the crops. Xenophon reports that this area was fertile and teaming with game;[2] today this section of the world is largely desert. Less than 20 percent of Iraq is cultivated, and more than half of the national income is from oil. In Babylon the silting of the canals became such a problem that the people there and their slaves spent most of their time cleaning them out and had little time for leisure. The silt was piled up along the banks; as these grew higher they were more susceptible to erosion by violent winds and rains, and the canals filled up more and more quickly.

At one time Ur was a seaport, but today it is 150 miles from the sea with its buildings buried under 25 feet of silt. The soil from the alluvial plain here has been washed out into the Persian Gulf.

The Yellow River of China bears that name because of the heavy load of silt which it carries.

Of all of the major river systems of the world the only one that has so far been able to withstand the abuse inflicted on it by man is the Nile; its headwaters have not been affected by human development and consequently it can continue to be relatively fertile.

Soil Erosion in the United States

Our own country has suffered a great deal from soil erosion. About one third of the land in the United States (600,000,000 acres) could be used for agriculture; about 400 million acres are in use, the other 200 million would require much more work and more careful erosion control techniques. If we add the grazing area to the land used for agriculture, we find that about 60 percent of the land in the United States is used directly for crops or for livestock.

In the United States it is estimated that 280 million acres have been severely damaged by erosion, and another seventy-seven and one-half million acres have been eroded to some extent. The Soil Conservation Service estimates that about 180 million acres suffer some erosion annually. This represents a loss in value of about $1 billion per year or 1.5 percent of all land values.

In the United States the early settlers found a virgin land with a deep soft forest topsoil. There streams ran clear, and there were few floods. Settlement was accompanied by a great deal of waste. The forests were cut in order to clear the land for crops. Much of the wood was burned; the ash was sold in Europe as potash – a practice which represented exportation of the mineral wealth of the country. In other places the tan bark was removed from hemlocks and the trees left to rot. In some sections of the south pines were "boxed" for turpentine; these trees were so weakened that many of them blew down or were destroyed.

In cutting the forests the early settlers removed the cover. Forest soils act as a sponge, absorbing the rain as it falls and gradually releasing it in the form of springs or in surface streams. As a consequence there is little real flooding in a forested area.

Forests Prevent Soil Erosion

The forest is extremely important in preventing soil erosion. One study near Zanesville, Ohio, showed scarcely any loss of soil in a 9-year period in a woodland, whereas there was substantial loss in adjacent areas. The pasture adjoining the woodland was found to be losing topsoil at the rate of one inch every 1,500 years, and the crop land which had practically no cover was losing soil at the rate of one inch every 9 years.[3]

We can understand the attitude of these early pioneers. They had to fell the forest in order to wrest a living from the soil. To them the forests represented a ruthless enemy. Not only was it necessary to fell the trees, the roots had to be dug out by hand before crops could be planted. Moreover, many people felt that the forests were the home of wild animals which threatened their safety and well-being. They were in error, for there are few if any authentic reports of attacks on man even by the larger carnivores, but the myth persisted and encouraged people to fell the forests.

Recognizing That Forests Need Protection

It was not long until the need for forest reserves became apparent. Already in 1799 Congress authorized the spending of $200,000 to buy reserves of live oak in South Carolina and Georgia. These trees were needed for our developing navy. Later other lands were acquired. The program, however well intentioned, proved to be a fiasco. People stole the timber from government lands, and the government itself sold oak timberland at the price of $1.25 an acre, buying back the oak at a price of $1.50 a cubic foot. In 1831 Congress prohibited the cutting of live oak and other trees on naval reservations, but this law was not enforced.

The Civil War alerted us to the need for forest preservation so that we might continue to have lumber. Wisconsin set up the first forestry study commission in 1867. New York stopped the sale of tax reverted lands in the Adirondacks in 1883 and used these lands as the basis for the Adirondack Forest Reserve.

One of the major problems was the practice in some states of buying forested areas, mortgaging them, cutting the timber, and then abandoning them because the land was no longer worth as much as the mortgage; a number of states had to pass laws prohibiting this practice.

Developing a Forestry Policy

Our national forestry policy developed between 1897 and 1919; Gifford Pinchot was active in its development. Beginning in 1919 we saw private organizations beginning to develop forest reserves. At the present time more than one fifth of the land area of the United States is privately owned forest land and most of the commercial forest land is in the hands of private individuals and corporations.[4]

Types of Erosion

The felling of our forests was a substantial factor in promoting soil erosion. Coupled with this were poor farming practices with little effort to reduce soil erosion. Perhaps the most widespread type of soil erosion is sheet erosion. As water passes over a smooth slope it flows in a sheet of more or less uniform depth. This sheet of water picks up soil particles and carries them along downhill. While this is the most common type of erosion and does the most damage it is rarely recognized because this type of erosion is uniform. Other types of erosion are rill washing, which results in the development of shoestring gullies, gully erosion, waterfall erosion, landslide or slip erosion, splash erosion from large raindrops, and erosion by waves in oceans and lakes.

Wind erosion is also a major factor – particularly in the Great Plains area, the Columbia River Plains, some parts of the Pacific Southwest, the Colorado Basin, the muck and sand area of the Great Lakes, and the sand areas of the Gulf and Atlantic seaboards. The most dangerous period for wind erosion is winter and early spring when there is little vegetation cover. This is coupled with the alternate freezing and thawing of the soil, which breaks up the clods and makes it easier for the wind to pick up dust and small particles.

The Soil Conservation Service has been reasonably successful in reducing the amount of damage to our land by suggesting ways in which individuals can reduce soil erosion. Wind erosion can be reduced by the planting of tree shelter belts; contour plowing is one way of protecting the soil. Terracing is another way of preventing soil erosion and also making the best use of water. Strip cropping is also successful in some areas; here closely sown crops alternate with strips of other crops.

Soil erosion is a problem not only on the land from which the soil is removed but also in places where it is deposited. Erosion is always accompanied by deposition, and in many cases the soil and silt are deposited in our streams. The streams become muddy, and many game fish are killed off.

In addition soil erosion contributes to the rapid filling of irrigation, flood control, and hydroelectric dams. Most large dams have a life expectancy of somewhere between 50 and 100 years because of the burden of silt which is deposited behind them. From the standpoint of flood control small dams in the headwaters of streams are much more successful than large dams, because the small dams hold the silt near the headwaters and also control runoff at this point.

Strip Mining

Another aspect of damage to the soil is the practice of strip mining. There was a time when most of our coal was gotten by sinking a shaft and digging the coal out from the seams in which it occurred. Rarely was a coal seam close enough to the surface to be removed by surface mining techniques. However with the development of newer and larger earth-moving machines it is possible to remove an overburden of 100 feet or more and then to remove the coal by strip mining techniques. Some of these machines are as high as a 21-story building and can scoop up 200 tons of earth and rock in a single scoop.[5]

Unfortunately in such procedures the topsoil is first removed, and the subsoil which has little capacity for supporting vegetation is piled on top of it. Moreover strip mining techniques inflict ugly scars on the landscape by producing alternate hills and gullies. Because these have little or no vegetation they quickly erode and scar.

It is estimated that 2 million acres of fish and wildlife habitat have been destroyed by strip mining procedures. At present about two thirds of the stripped areas are unreclaimed; about one third have been reclaimed. Of the areas that have been reclaimed 46 percent have been reclaimed naturally, 40 percent voluntarily by strip mining companies, 11 percent by strip miners under legal compulsion, and 3 percent by federal, state, and local governments.

In some places strip miners have been required to post bond as a guarantee that they will restore the land to a state in which nature can quickly reclaim it. However in some cases the bond has been much less than the cost of restoring the land, and the result has been that irresponsible companies have forfeited the bond without restoring the land as required.

Some farsighted coal companies have developed programs of land restoration which have not only removed the strip mining scars from the landscape but have provided additional recreational and pasture lands. By smoothing over the spoil banks they have provided a gently undulating landscape with artificial lakes in the low places, which have been stocked with fish. In addition it has been found that certain grasses of considerable nutritional value to cattle can be grown very quickly on stripped over land.

Thermal Pollution

A relatively new area of pollution is that of thermal pollution. We are only beginning to realize that slight changes in the temperature of surface waters may substantially alter the ecological balance. We have known for a long time that certain fish species prefer colder waters, but we have not realized how slight changes in temperature will lead to their disappearance.

The chief sources of thermal pollution have been electric power plants. Increasing demands for electricity have resulted in thermal loading of the aquatic environment, which is becoming common to all highly industrialized regions of the world. Both conventional and nuclear power plants require large quantities of water for cooling. In the United States 90 billion gallons of water were required in 1958 for cooling purposes in steam generating utilities using fossil fuel as their source.

It is estimated that by 1980 conventional power plants will use between one fifth and one sixth of the total fresh water runoff in the United States. Discounting flood flows which occur about one third of the year and account for two thirds of the total runoff, the steam electric industry will require from 40 to 50 percent of the total fresh water runoff for cooling purposes the remaining two thirds of the year.[6]

Added to this are the needs of nuclear power plants; these require even larger quantities of water for cooling purposes.

Thermal pollution can be reduced and even prevented by building cooling lakes and cooling towers. However these are expensive and will add to the cost of electricity. Yet if we want to preserve our environment, we shall have to be prepared to pay the price. We shall have to insist that cooling towers be provided so that the water discharged into our surface streams does not increase their temperature to any appreciable degree, and we shall have to be willing to pay more for the electricity which we use.

We also need research into uses that may be made of this waste heat. The heat represents energy, and it ought to be possible to find some use for this energy. Studies have been made on the use of this heated water for agricultural purposes, for keeping open frozen river channels, and for removing ice and snow during the winter.

Many of these problems can be overcome, but overcoming them will cost money; this will have to be reflected in increasing cost paid by the general public for the electric power which they demand.

Noise Pollution

Still another area of pollution is that of noise pollution. Sound is measured in decibels; the amount of noise in an average home is about 40 decibels. A heavy truck produces 90 decibels, a subway train 95 decibels. Construction noises of hammers and compressors produce about 110 decibels of sound. An overhead aircraft at 500 feet produces 115 decibels of sound, a rock and roll band at its peak about 120, and a large pneumatic riveter at three feet about 125.[7]

A particular problem today is the development of supersonic transports. It is believed that when these are in full operation sometime in the late 1970s, about 65 million people in the United States will be exposed to an average of 10 sonic booms per day. There is some feeling that it will be necessary to restrict the supersonic transports to travel over water or over sparsely populated land areas and that they may be permitted over populated areas only very occasionally.[8]

Sonic booms are damaging prehistoric buildings and rock formations in the national parks and monuments in the southwestern United States. The Transportation Department has commissioned a natural environmental panel to study the effects of sonic boom on national parks, and data recorders are being placed in Yellowstone, Bryce, Yosemite, and Mesa Verde to record the frequency and intensity of booms. It is feared that undetected damage may already have occurred in remote park areas.

On October 12, 1966, shortly after three exceptionally sharp booms, approximately 10 to 15 tons of dirt and rock were found to have fallen from a formation in one of the parks. The possibility of new collapses in such areas as the Canyon de Chelley National Monument in northeastern Arizona is particularly disturbing because many of the park ruins which could be destroyed have never been fully excavated by archaeologists.[9]

Other city and industrial noises undoubtedly have their effects, and decline in hearing acuity is inevitable. Furthermore the effect of a constant stream of auditory stimuli cannot be ignored.

In this connection a reduction in wooded areas in and around our cities and the cutting of shade trees undoubtedly has its effect on noise pollution. Trees and shrubs serve as effective sound barriers. Properly planted, both can reduce street noises for the homeowner. Trees with spreading high crowns can lessen the intensity of violent noise from aircraft. We probably need to study the efficiency of different shrubs and cultivated trees in lessening urban noise.[10]

Solid Wastes

Still another aspect of pollution is that of solid wastes. In 1920 an average of 2.75 pounds of waste were produced each day by each individual in the United States. Today the quantity of waste produced is 5.3 pounds per person, and by 1980 it is estimated that this will rise to 8 pounds per person. One year's rubbish from 10,000 people covers an acre of ground to the depth of 10 feet. In one year Americans throw away 48 billion cans, 26 billion bottles, 430 million tons of paper, 4 million tons of plastic, and 100 million tires which weigh almost a million tons.

The cost of disposing of these wastes, totally about 3.5 billion tons per year, is some 4.5 billion dollars and is exceeded only by expenditures for schools and roads.

We live in a "throwaway" economy, and the

development of these throwaway items has aggravated our problem of solid wastes. The development of such a simple thing as disposable syringes in hospitals has created a real problem. We used to wash and sterilize our syringes and reuse them; but today we throw them away, and since they are a synthetic material we do not know how to get rid of them effectively.

Tin cans are creating an increasing problem. There was a time when soft drinks were bottled in glass containers which were reusable. The individual had to return them in order to get back his deposit. This was an inconvenience and annoyance, and the old deposit bottle was succeeded by the tin can which, while it stayed longer, eventually rusted away when it was discarded. Tin cans were eventually replaced by aluminum cans which have greater staying powers. Now, however, we are in the "no deposit-no return" glass bottle stage. Someone has suggested that these will undoubtedly be around in vast numbers to bore future archaeologists who some day dig in the mounds representing our cities.

The problem of solid wastes is world wide; Thor Heyerdahl sailed in the Atlantic in a reed boat and reported large expanses of floating wastes in the middle of the ocean. The Quinault Indians of Washington recently barred campers from 25 miles of ocean front because of what they regarded as an offensive litter problem.

Disposing of solid wastes is not easy. If we incinerate them, we contribute to air pollution; many communities have barred open burning of trash and leaves, and while municipal incinerators are better than open burning by individuals they too contribute to air pollution.

At present sanitary landfills probably provide the best solution. In this practice the wastes are buried each day in a trench, and the trench is covered daily to discourage rats and vermin. Sanitary landfills require land, however. It is estimated that one acre per year is required for every 10,000 people. Moreover, they must be carefully located with regard to underground drainage patterns so that seepage from them does not contaminate the aquifer.

Research is being carried on to determine whether solid wastes can be compressed into some useful product; it has been suggested that they be used as a cheap building material.

Abandoned Cars

Another problem is that of abandoned automobiles. Throughout the United States there are 2,500 automobiles abandoned each day. In terms of environmental pollution this number of cars uses up 6 acres of otherwise attractive and usable land each day. Besides contributing to the "uglification" of our rural and city areas the abandoned cars, their broken glass, and their jagged metal edges present a serious hazard to small children who cannot resist their appeal as a playground.

Moreover the abandoned car is a tragic waste of valuable scrap metal resources.

A number of suggestions have been made to solve the problem posed by these cars. We shall probably have to provide financial and other aid to local governments, helping them to overcome problems of moving unprepared scrap metal to the scrap processor. We shall also need government cooperation in changing title laws to speed up the processing of the abandoned cars, and we shall need government and industry cooperation in providing expanding markets for processed iron and steel scrap.

NOTES TO CHAPTER 6

1. Raymond F. Dasmann, *Environmental Conservation* (New York: Wiley, 1968), pp. 70 f.
2. Edward Hyams, *Soil and Civilization* (London: Thames and Hudson, 1952), p. 59.
3. B. D. Blakely, J. J. Coyle, and J. G. Steele, *Soil: the Yearbook of Agriculture for 1957* (Washington: U. S. Government Printing Office, 1957), p. 290.
4. Dasmann, p. 189.
5. Theodore B. Plair, *Outdoors U. S. A.: the Yearbook of Agriculture for 1967* (Washington: U. S. Government Printing Office, 1967), pp. 361-3.
6. *BioScience,* 17 (1967): 698.
7. Bruce L. Welch, "Physiological Effects of Audible Sound," *Science,* 166 (1969): 533.
8. Karl D. Kryter, "Sonic Booms from Supersonic Transport," *Science,* 163 (1969): 359.
9. *Science,* 161 (1968): 343.
10. H. L. Li, "Urban Botany: Need for a New Science," *BioScience,* 19 (1969): 883.

Chapter 7

Threatened Species and Natural Areas

Another real problem of environmental deterioration is the extinction of species. It was the extinction of the passenger pigeon and the near extinction of the buffalo which began a conservation movement in the United States.

The original population of the passenger pigeon in the United States is believed to have been somewhere between 3 billion and 5 billion. Hawks and other predators indeed preyed on them but seemed to have had little effect on their numbers. One factor contributing to their extinction was the cutting of the mast trees, the beeches, and the oaks, for these birds were essentially birds of the deciduous forests living on the nuts and acorns produced there. There is little doubt that the cutting of the forests in order to provide croplands was a serious blow.

Man also contributed by trapping them and by interfering with their breeding. He took the squabs or young pigeons from the nest because he found them the best eating. Weather played a part in the form of a number of severe late spring storms at a time when the birds were on the verge of extinction.

The last of the passenger pigeons, Martha, died in the Cincinnati Zoo in 1914.

Extinct Species

It is worth noting that over 20 species of birds have become extinct in the United States since the coming of the white man. In addition to the passenger pigeon, which became extinct for all practical purposes in 1898, although the last pigeon died only in 1914, the Carolina paroquet became extinct in 1904. The Townshend bunting became extinct in 1832, the Gosse macaw in 1800, the black-capped petrel in 1912, the Michigan grayling in 1930, and the heath hen in 1931.

Most of the organisms that have become extinct in recent times have been birds, but some mammals too have disappeared. The Stellar seacow died out in 1768, the sea mink in 1860, the California grizzly in 1900, and the Arizona elk in 1901.

Organisms continue to become extinct as the years pass. Probably the most recent mammal to become extinct was the Mexican grizzly bear; according to the World Wildlife Fund there have been no recent sightings of this species.

At present there are 89 forms on the "endangered species list" compared to 78 that were there in 1967. Since the turn of the century an average of one species a year has quietly made its exit somewhere in the world, and it is believed that the total number of species endangered on the worldwide level is about 275 mammals and over 300 species of birds. Among the presently threatened species are the California condor, the giant otter of South America, the western giant eland, the tiger, the polar bear, the Tasmanian wolf, and the giant sable antelope.

Causes of Extinction

What brings about the extinction of a species? Some of course are the victims of biological eclipse, a natural extinction of species. But this is usually a slow-moving process. It may be, however, that the California condor, the ivory-billed woodpecker, the Kirtland warbler, and the manatee are threatened by this form of extinction. While this cannot be prevented, man can help such threatened organisms by easing the pressures. Protection will help, and they need not become extinct in our lifetime.

In this connection there is such a thing as an extinction threshold. An organism may actually become extinct before the last member of the species dies. When hunting and trapping of the passenger pigeon was forbidden there were still several thousand birds alive. Yet within several decades the species died out. Widely scattered over the continent, the birds probably had difficulty in finding mates. Moreover, because breeding communities often included several hundred birds in a single tree, it may be that the passenger pigeon was unable to breed when its numbers were so low. It seems that in many cases it is sufficient for extinction to reduce the population below a certain minimum number which the species seems to require to survive. Population of several species such as the whooping crane and Philippi's seal may be doomed in spite of desperate efforts to preserve them.

The inroads of settlement also play a part. Animals may be deliberately destroyed under the guise of disease control, as was the case with the Florida deer. Sometimes man kills off animals to cut off the food supply of other human beings whom he regards as his enemy; this was the case with the bison. It was thought that the plains Indians would be pacified if their food supply, the buffalo, was killed off, and this became deliberate army policy.

Hunting for sport and trophy is another factor of extinction. Unfortunately the largest, the best, and the rarest are usually hunted for trophies. The desire for possession is strong, and most egg collections are made just for the joy of possession. The rarer the organism the more valuable the trophy and the more valuable also its eggs.

Animals and birds may also be pursued for plumage, fur, and other products. Philippi's fur seals of Juan Fernandez Island off the Chilean coast were three million strong in the late 1700s. Only 50 have survived man's greed for pelts.

Ignorance and maliciousness may also lead to wanton destruction of animals. Some people shoot any live or unusual animal just to see it or merely to try to hit it. Along ocean cliffs accessible from highways people have maliciously thrown objects or fired guns at nesting sea birds or even at sea otters.

Elk were formerly hunted for their teeth, and the rest of the animal was left to rot. Whales were hunted ruthlessly and wastefully. Kingfishers, owls, and hawks are hunted because, being predators, they are incorrectly regarded as threats to man's welfare.

Poison campaigns have destroyed not only birds and mammals thought to be predators but many birds and animals considered by everyone to be harmless.

Introduction of Exotic Species

Man also does a great deal of damage by the exotics he introduces. Sometimes this is done deliberately, sometimes accidentally. The introduction of exotic species is a particular hazard on islands where rats, pigs, goats, and sheep devour the vegetation and destroy the habitat of native species.

Past Successes in Saving Threatened Forms

It is sometimes possible to reverse the trend toward extinction. The bison was an endangered species when President Theodore Roosevelt and Gifford Pinchot became concerned; today it has apparently been saved for posterity. In 1950 there were only 30 nene geese, a Hawaiian species that is the state bird of our 50th state. At one time flocks numbered 25,000, but they were reduced by overhunting, by wild dogs, and by wild pigs. Today there are 50 wild nenes known and 150 in captivity. A similar situation exists with the whooping crane. We are not yet sure that we have succeeded in saving it, but the number of whooping cranes has increased slightly, and we are hopeful that we have saved it.

It should be pointed out that some species have actually increased in numbers. There is no doubt that both the white-tailed and the mule deer have increased since settlement days. Elk and black bear are also on the increase. Those animals whose numbers have increased since the advent of settlement are generally animals which thrive on second growth; the white-tailed deer is a good example of this. The white-tailed deer is not an animal of the forest but of forest edges.

What to Do Today

What can be done to protect endangered species? One of the most important things is the provision of habitat. Many animals are threatened because we are encroaching on their territories. This may be a major factor in the decline of the California condor. We need to acquire areas where animals are protected from human encroachment.

Many well-meaning people advocate the provision of artificial feeding programs for desired species, but this is not a good solution; in many cases these programs do far more harm than they do good. Far more important is the provision of additional habitat where animals can feed on natural foods and where they can breed unmolested and undisturbed.

In this connection we might note that the large mammals of Africa are threatened by the encroachment of civilization. Burgeoning populations have created the need for additional land for agricultural purposes, and the areas once available for wildlife are being fenced and farmed.

The World Wildlife Fund has been active in acquiring land which serves to provide the desired habitat for threatened and endangered species.

The provision of additional habitat for threatened species is particularly important because of man's alteration of the environment in his own interest. He clears forests and replaces them with monocultures such as coffee plantations and cotton fields; in this way entire communities of animals are left without food and shelter. He fills in swamps and marshes and thus destroys the breeding and nesting ground of marsh birds. He builds dams and changes the courses of rivers, making the salmon's migration from sea to inland spawning grounds increasingly difficult.

Man also leaves a trail of poison behind him. He seeks to eliminate supposedly competitive carnivores not only by intensive hunting but by leaving poisoned sheep and cattle carcasses hoping the carnivores will feed on them. Often these poisoned carcasses are eaten by scavengers; this may be another contributing factor in the decline of forms such as the California condor.

Another important helpful procedure is restricting by law the consumption of wild animals and the wild animal trade. In the early 1900s the Audubon Society found its game wardens fighting a losing battle to save the American egret, a bird common in the south; its large white feathers were prized for ladies' hats. When the Federal Government outlawed the trade in egret feathers, egret poaching came to a halt.

It is hoped that laws similar to the New York City ordinance forbidding the sale of alligator hides will protect this animal, which is threatened today. At one time there were a million alligators in the Everglades National Park; today the entire alligator population in the United States has dropped to 200,000, and it is declining rapidly in the face of heavy poaching and marsh drainage.

A recently passed federal law protects both native and endangered foreign species, forbidding the importation of live animals as pets and the importation of hides and furs for coats. The Secretary of the Interior will maintain a list of fish and wildlife threatened with extinction anywhere in the world. Federal inspectors at ports of entry, warehouses, or retail stores may confiscate imported live specimens or any imported skins, coats, or manufactured items made from a species on the list. Federal law now prohibits the interstate commerce of birds, fish, or mammals poached in any state whether or not they are rare. The new law extends this protection to reptiles, amphibians, mollusks, and crustaceans taken in violation of federal, state, and foreign laws and to fish taken illegally in a foreign country. The law exempts the importation of endangered species for zoological, educational, scientific, or propagation purposes.[2]

Another procedure that may be helpful is a program of captive propagation of threatened species. Since 1966 the Bureau of Sport Fisheries and Wildlife has maintained a captive propagation program at its Patuxent Research Refuge in Maryland. The Bureau is experimenting with breeding stocks of the Hawaiian nene goose, the Aleutian Canada goose, the southern bald eagle, the whooping crane, and the masked bobwhite. A number of zoos are also emphasizing captive propagation rather than the further collection of endangered foreign species.

It is hoped that when these animals do breed at least some of them will be transferred not to zoological parks but to their native habitat. It may be that this will not always be possible, since in many cases artificially raised animals cannot survive in the wild.

Saving the Galapagos Tortoises

A program of this nature is being carried on in the Galapagos Islands, where the Darwin Research Station is attempting to propagate the various species of Galapagos tortoises which are threatened with extinction. The plan there is to propagate them at the research station and then transfer them to the island from which their parents were taken.

Natural Area Preservation

Closely related to the problem of endangered species is that of preserving relatively undisturbed and therefore relatively natural areas. Our national parks have provided this to some extent, but they are in danger of being "loved to death" by burgeoning populations. A generation ago such a program was unnecessary because there was a great deal of undeveloped land and because almost every farm had its woodlot which provided a relatively natural and undisturbed area.

Today the need for setting aside these undisturbed areas is great, and this need cannot be met by the Federal Government in its national parks and national forests or by the state governments in their state parks and forests. We particularly need smaller areas close to our centers of large urban populations.

The Need for Natural Areas

Prince Bernhard of the Netherlands recently commented on a conservation report by saying that it was "incomplete in one critical respect:" it failed to deal with "the role of private institutions that are already in a position to *act* and their need for . . . financial support. These bodies require money—big money—primarily to buy land to set aside permanently for conservation uses. It is often a costly race against property speculators; unless we can win, all the publicity and goodwill are of little avail. Conservation is an infant among charities. It cannot grow to effective maturity until it obtains adequate financial support. Only upon this second awakening will conservation have a fighting chance."[3]

These natural areas are being disturbed by a variety of factors. As our growing populations move into metropolitan areas and these extend their boundaries, urbanization is a major factor. Another factor is that developing technology makes possible the exploitation of areas which in previous years could not be economically exploited. Often some of the most desirable areas from a conservation standpoint fall into this category; formerly they were bypassed because they were too rugged to be used. Now, with the development of our giant earthmoving machinery, we can push over trees, level hills, remove boulders, and alter the contours of the land.

Another factor has been the building of our extensive superhighway network which has taken over a great deal of the landscape. A four-lane highway consumes up to 50 acres of land per mile, preferably the most level and most strategic land available.[4]

Industrial uses of suburban areas have also increased.

Natural areas have also been drained or leveled or forests felled in order to provide additional land for agriculture—and that at a time when agricultural surpluses continue to mount.

Like air and water, open space is necessary for the survival of human beings, even though we are not yet able to estimate how much is needed. The more the pressure for development increases, the more we need to provide a margin of safety in open space. In earlier times, with fewer people and slower growth of our populations, there was little danger of pollution or urban sprawl, and consequently little real impact on the environment. Now we need to ask how we can provide enough open space to live properly. The continual overuse of land by subdividers may be detrimental to the public interest in the long run. It is inevitable that we will continue to suffer overcrowding, erosion, flash floods, pollution, traffic jams, and loss of trees unless we take action to insure the preservation of green belts and natural areas.

Two kinds of resources are involved in conservation: essential resources such as food, minerals, water, and living space; and desirable resources such as wildlife, play space, and walking space.

The Need for Zoning

Also to be considered is the desirability from an economic standpoint of leaving certain types of land undisturbed. Certain types of land do not lend themselves for homesites or for industry; people need to be protected against their own foolishness by restricting the type of development permitted on these sites through zoning laws.

One type of land that should be left undeveloped is the floodplain area. Such areas are often level and inviting for building purposes, but they flood regularly and the cost of protecting them can be exhorbitant.

Fragile sites like deep slopes, deserts, hilltops, and lake fronts need to be protected, although they too can apparently be built on. Land contour should be followed in building, for natural drainage can often save the cost of storm sewers. Forest remnants, steep woods, and landmark trees should be retained since vegetation keeps homes cooler in summer and warmer in winter.

These areas, which are not really suitable for homesites or for industry, provide green spaces and natural areas for the community. In the long run the cost of developing these, when added to the reduction in land values because there are no green spaces, is greater than the value of the land itself.

Open spaces are needed to relieve the pressures generated by close living in our cities. While we cannot prove that such is a major factor with man, laboratory studies of animals indicate that all sorts of undesirable characteristics develop when animals are crowded together, and there is every reason to believe that some of the stresses we note with human beings are due to crowding. Open spaces and undisturbed areas, where man can get away from his asphalt jungles, assist in reducing some of the tensions of crowded living.

Living Museums

Undisturbed areas can also serve as living museums. We need to know what the land was like when the first pioneers came to it and before man came and altered the landscape. Such living museums will give us a greater appreciation of the conditions the pioneers faced and under which they lived.

Other areas are needed to preserve both plants and animals. Once a natural area has been bulldozed it will take generations for plants to return if indeed this ever occurs. There is real beauty in our forested areas; a man cannot help but be impressed by an aged giant of the forest or by beautiful wild-flower displays.

Natural areas are often desirable as habitat for birds and wildlife. Even though they are found close to our metropolitan areas and are not likely to be endangered species, birds and animals contribute to the joy of living. Most children and adults need the thrill of seeing a deer or a fox or a raccoon.

Unanticipated Needs

Natural areas are desirable as a resource against needs which are unanticipated at present. Who would have believed 50 years ago that one of our most important medicinals would be derived from a lowly mold, *Penicillium notatum*. We need to protect natural areas so that we do not lose any potentially desirable species.

The development of new varieties of food plants must be more or less continuous. Frequently, as soon as research scientists breed a variety which resists disease, the microorganism develops a different race or strain of the pathogen or even a different pathogen to attack it. So the plant breeder has a problem just to keep one jump ahead.

For this purpose undisturbed natural areas are most important. Here the breeder searches for native plants having desirable characteristics. When natural areas are destroyed, his problem becomes more serious. Every effort must be made to preserve a suitable number of natural areas over the world. Every year of delay in a program of this sort means that potentially important plants bearing genes for resistance or other valuable characteristics are becoming extinct.[5]

Some organisms, poisonous under certain conditions, provide us with desirable drugs under other conditions. Ergot is a good example of this. The fungus grows as long, hard, purple masses replacing some of the grains in a maturing head of rye. It was often included in low-cost bread until its poisonous nature was discovered. A toxic alkaloid in the fungus induces abortion and gangrene. In larger quantities it brings about uncontrollable hysterical laughter and may even lead to death. Epidemics of ergot poisoning swept over France nine times in the 17th century and eight times in the 18th century and thousands died. Epidemics have also occurred in the 20th century.

Yet it is possible to extract from ergot a drug which has wide medicinal uses. The fact that in some doses it appears to be toxic does not mean it is without value to man. There may be similar substances which deserve further study and research.

Areas for Research

Also to be considered is the need for natural areas as places where we can study crop improvement. It has often been erroneously thought that if some of the South American, African, and Asian countries would use our seed and our methods there would be plenty of food to go around. Unfortunately the procedures that work in Europe and North America are quite apt to fail in other places. For that matter procedures that are suitable in southern Wisconsin may not be suitable in northern Wisconsin. Research is necessary in order to learn the application of basic knowledge to the problems of particular localities, and for this purpose we need relatively undisturbed areas.[6]

A Need Recognized

Recognition of the importance of undisturbed areas is gradually dawning. A good example is the Everglades jetport controversy which seems to have been resolved in favor of protecting the area. The Environmental Studies Board of the National Academy of Sciences presented a very carefully prepared report on the problems of commercial development near the Everglades and concluded that the development of the jetport could lead to disastrous consequences, unless industrial and residential development was kept at a minimum and adequate water resource management was practiced—conditions which are unlikely to obtain.[7] For that reason it appears that the project will be abandoned.

Another example is the changes that were made in the Red River Dam in eastern Kentucky. The Red River rises in the Cumberland plateau and flows westward into the Kentucky River. At the western end of the plateau its north fork has cut a gorge of rare beauty which biologists and geologists consider the best "outdoor laboratory" in eastern Kentucky. The Corps of Engineers proposed several years ago that a dam be built here, and its construction was authorized in 1962. Funds for the dam were appropriated in 1967 and in 1968. Concern over the damage that the dam might do to this unusual area was expressed by a number of professional biologists, who carried their plea to

Congress. State officials also became interested and involved with the result that the originally proposed site for the dam has been abandoned.[8]

NOTES TO CHAPTER 7

1. Mark W. Oberle, "Endangered Species: Congress Curbs International Trade in Rare Animals," *Science,* 167 (1970): 152-4.
2. Ibid.
3. *Time,* 95 No. 8 (Feb. 23, 1970): 4.
4. *BioScience,* 17 (1967): 874.
5. A. J. Riker, "Plant Pathology and Human Welfare," *Science,* 152 (1966): 1029.
6. Ibid., 1030.
7. Marti Mueller, "Everglades Jetport: Academy Prepares a Model," *Science,* 166 (1969): 202.
8. Mary Eugenia Wharton, "Red River Dam Controversy," *American Assn. for the Advancement of Science Bulletin,* 14 (June 1969), 1f.

Chapter 8

Radiation Damage

Another area of damage to the environment is that brought about by radiation. The total number of nuclides, both radioactive and nonradioactive, is in excess of 1,300. Of these, 279 stable forms have been identified; these create no problems. Over 1,000 radioactive nuclides have been identified—51 of these occur naturally and another 100 are now being produced artificially in radioactive reactors.

Types of Particles

There are three kinds of particles which must be considered. Alpha particles are high velocity helium nuclei stripped of atomic electrons. They travel relatively slowly and over a short course because of their relatively great weight. They have a positive charge and consequently may be involved in a great many interactions. Close contact with alpha particles may dislodge an electron and cause ionization. However, they are not too hazardous to man because of low penetrability and of the protection afforded by a layer of keratin. In general they are regarded as not too important externally, but they may be harmful if ingested or inhaled in a nuclear holocaust.

Beta radiation consists of a series of high-energy electrons. If a beta particle comes close to another electron it may dislodge the other electron. These are very light, they are easily deflected, and pursue a zigzag course. Penetration is poor and the only damage likely is possible damage to the skin by external application.

Gamma rays and x-rays are similar; they differ only in their origin. They are very energetic and behave both as rays and as particles. They carry substantial quantities of energy and can cause ionization by a direct hit but have no effect if they only come close. It is with gamma rays and x-rays that we are most concerned in considering environmental damage. The source of gamma rays is intranuclear; that of x-rays is extranuclear. Gamma rays are usually of shorter wave length than x-rays and accordingly have higher energy.[1]

Measuring Units

There are a number of units of measurement in studying ionizing radiation. One unit, the roentgen, is used in monitoring work, and it is valuable in measuring x-rays or gamma rays. The rep (roentgen equivalent physical) and the rad (radiation absorbed dose) are defined in terms of absorbed energy and are equivalent to the roentgen. A rep is the amount of radiation which causes the transfer to a tissue of the same amount of energy as one roentgen. The rad is equal to 100 ergs of energy absorbed per gram of any absorber.

Another unit which is used is the rem (roentgen equivalent man). It is obtained by multiplying the rad times a factor called the rbe (relative biological effectiveness). The rem does the same amount of damage via radiation as the amount produced by exposure to one roentgen of 200 to 250 kilovolt x-rays. Rbe factors vary from unity for x-rays, most gamma rays, and lower energy beta particles to around 10 for fast neutrons and around 20 for one-Mev alpha particles, depending on the particular biological end point studied.[2]

Background Radiation

Radiation is a part of the natural environment. We are all exposed to radiation from cosmic rays, from radioactive rocks, and from other radioactive materials in the environment. This constitutes background radiation, and we have no control over it. It has been suggested that this may be responsible for the natural mutation rate in man, though this still remains to be demonstrated.

Artificial Radiation

The factor over which we have some control is man-made radiation. The chief source of this is medical x-rays. In 1956 geneticists on a National Academy of Sciences committee recommended that the contribution of man-made radiation to the human body not exceed 10 rems per generation (30 years). At that time it was estimated that exposure to some medical uses of radiation accounted for about one half of this value. The remaining 5 rems divided by 30 years gives a figure of about 0.17 rems per year.[3] Since that time, some have felt even this figure too high and have advised reducing it by 50 percent.

Just what are the hazards of radiation? These must be divided into at least three categories. There is the short-range damage to the individual, there is the long-range damage to the individual, and finally changes in hereditary factors which may be passed on and may affect the future of the race.

Short-range Damage to the Individual

It is believed that a dose of between 400 and 600 rems is generally fatal. Such exposures might be experienced during war; this was the cause of a large number of deaths at Hiroshima and Nagasaki. Accidents in plants using nuclear materials may expose individuals to fatal doses. In July 1964 a worker in a uranium processing plant was exposed to 8,800 rads and died in 49 hours. In 1967 Edward Czwalda received 600 rems. He was fortunate in that he had an identical twin, and it was possible to save his life by transplanting bone marrow from his twin.

Exposure to an atomic explosive may result in burns and blast damage to the individual, but these injuries are no different from those of other explosives.

Individuals exposed to a less than fatal amount of radiation may suffer radiation sickness. In this case the hair falls out, and the individual experiences considerable nausea. There is damage to the epithelium of the mouth, the stomach, and the intestines. In addition there is blood cell damage which results in anemia and in inability to combat even minor infections. Within an hour after 25 rems have been received a significant drop in the white cells in man can be detected. The blood picture with regard to red cells, which have a longer lifetime in the human system, shows a similar depression after radiation exposure but the depression begins after about a week.[4]

Long-time Damage to the Individual

Long-range radiation damage results chiefly in an increase in the amount of cancer and leukemia. Mortality from leukemia and other forms of cancer is about 40 percent higher among children exposed to diagnostic x-rays *in utero* than among children not so exposed. In addition a study showed an increase of 10 to 30 percent of cancer, primarily leukemia and cancer of the central nervous system, in children whose mothers were irradiated during pregnancy. Studies of mongoloid children have indicated that mothers of such children have been exposed to substantially higher amounts of radiation over a long period of time prior to pregnancy than the mothers of non-mongoloid children.[5]

Among Japanese survivors of the atomic bomb only leukemia and thyroid cancer have been found to be radiation induced. The evidence pertaining to cancer of the breast or lung is still very much in doubt.[6]

Abnormalities were observed in Japan among children born of women who were exposed to the atomic bomb while pregnant. The effect was primarily among children of women who were exposed within 15 weeks of their last menstrual period. Of the individuals examined 56 in this category were born of mothers who had been within 1,800 meters of the hypocenter. A head circumference two or more standard deviations below the mean for age and sex was expected in 2.5 percent of the group but actually observed in 23 or 41 percent, of whom 9 were mentally retarded. Among the 105 individuals exposed *in utero* more than 15 weeks after the mother's last menstrual period, small head circumference was expected in 2.6 but observed in 6, of whom 2 were mentally retarded.[7]

Some observers believe that there is a dosage below which no medically significant damage in humans has been observed and that this is somewhere between 50 and 100 rads. Others believe that there is considerable evidence indicating a linear response and absence of a threshold for the development of cancer and leukemia, because there is some evidence of the development of these at total doses below one roentgen.[8]

After correlation for birth order and other variables the average cancer mortality seems to be about 40 percent higher for children who have been x-rayed *in utero* than for those who were not. The rate is also higher for those irradiated during the first 6 months than the last 3 months. Indeed it is suggested that typical doses from background radiation which would amount to between 75 and 150 milleroentgens (a milleroentgen is 0.001 roentgen) in 9 months may be responsible for some 5 to 10 percent of all cases of childhood cancer and leukemia.[9]

There is also evidence of damage done to individuals through a shortening of life. Doctors who have no contact with radiation have a life expectancy of 65.7 years. Dermatologists and urologists who have some contact with radiation have a life expectancy of 63.3 years and radiologists a life expectancy of 60.5 years.

Long-range Species Damage

The third area of concern is that of mutation, and this is of great concern because it affects future generations. Damage to the individual ceases to be of concern when he dies; however mutations, which are passed on, continue in the human stock in future generations. The U. N. estimates that about 6 percent of live-born infants have visible defects of genetic origin. Most of these are caused by mutations which have occurred in past generations or through the recombination of genes present in some way in the human stock. About 1 percent of all live births show known gross chromosomal aberrations such as mongolism; these seem to be due to current radiation and chemical damage. Since most individuals so affected are sterile or have low reproductive capacity, these defects are rarely transmitted to subsequent generations.

The other genetic abnormality and the one which gives greatest concern involves point mutations, that is, single gene alterations that give rise to harmful effects with a known inheritance pattern. One percent of all live births are so affected.[10]

Reducing Exposure to X-Rays

It is this aspect of radiation about which we are most concerned. We cannot prevent mutations which occur as a result of background radiation, and the number of these is substantial. However we can reduce the amount of damage done by exposure to artificial radiation.

Among the worst offenders in this respect were the shoe x-ray machines, actually fluoroscopes. They gave off large quantities of radiation, and the benefits from their use were minimal; in most cases they were an advertising gimmick to sell shoes. In most states their use is now illegal.

We shall also have to reduce as much as possible

the use of medical x-rays. In every case the question "Is this x-ray necessary?" is in order. X-ray is a valuable tool for diagnosis and for therapy. Its use in setting bones is well known, and certainly it is too valuable to be discontinued. We also use x-rays for mass screening in the detection of tuberculosis. With the x-ray machines we have today the average chest x-ray involves about 45 milleroentgens. Wherever possible we probably ought to substitute the tuberculin test to diagnose tuberculosis; however there are some situations in which chest x-rays are still advisable.

Dental x-rays involve larger quantities of radiation; the average dental x-ray involves exposure to 1,138 milleroentgens of radiation. However very little of this reaches the gonads, and for that reason they do not seem to be too important as a radiation hazard.

Pelvic x-rays are particularly hazardous because much of the radiation is likely to reach the gonads. Every effort should be made to protect the gonads when pelvic x-rays are necessary. Because x-rays are particularly hazardous to embryos and fetuses, pelvic radiation of married women should be limited to the first 10 days of the menstrual cycle.

Radiation is widely used in therapy. Its use in treating skin disorders probably should be discontinued. It is also valuable in slowing down the progress of cancer. Since cancer is primarily a disease of the post-reproductive years its use in cancer therapy seems to pose little hazard. Even in those cancer patients who are still in the reproductive years the value of radiation in therapy outweighs the possible mutation hazards.

Nuclear War Hazards

It is obvious that we need to make every effort to prevent nuclear war. The bombs dropped on Japan were very small compared with the bombs in the arsenals of the nuclear nations today; yet the Hiroshima bomb killed about 100,000 people. A nuclear holocaust is unthinkable from a scientific standpoint. In such a case the fortunate individuals will not be the survivors but those whose lives are snuffed out immediately by an atomic blast. The survivors will find it necessary to grub in the ground for bare necessities. Moreover many of them will experience serious injuries, and it will be possible to treat only a limited number of individuals.

In addition, exposure to radiation in the nuclear war will add to our load of lethal, semilethal, and harmful mutations.

There is every reason for pessimism so far as the threat of a nuclear war is concerned, and scientists who know the hazards of such a holocaust are among the most pessimistic. Several years ago C. P. Snow predicted that if the Western World did not disarm unilaterally we would be involved in a nuclear holocaust by 1970.

Only the Christian can afford to be an optimist. He knows that the history of the world is being determined not in Washington, Moscow, and Peking, but in heaven. He sees the hand of God in the history of the past holding this protective hand over His Christians. He is confident that God loves him and that God will protect him.

Working for Peace

At the same time the Christian must work for peace and pray for peace. He is God's instrument in bringing about His will on earth. The Christian cannot stand with folded hands and permit the outbreak of a nuclear war. He must do everything possible to reduce the tensions which exist on an international level.

Fallout

There is real danger from fallout following the testing of nuclear weapons. We do not know how much damage has resulted from the bomb tests of the past years, but any damage is undesirable. It is to be hoped that the present moratorium on weapon testing will continue so that damage to future generations by fallout be held to a minimum.

Radioactive Wastes

Another source of radiation damage are the radioactive wastes produced by nuclear power plants. These must be disposed of, and the common practice today is either to pump them deep into the earth or to pack them in drums which are dumped into the oceans.

In any case there is hazard of leakage. It is possible that the radioactive wastes pumped into the earth may reach the aquifer, and it is possible that the drums that are dumped into the oceans may leak. We need to carry on research to determine better ways of disposing of these radioactive wastes.

Nuclear Power Plants

Present nuclear power plants employ fission methods. It has been suggested that fusion methods may enable us to produce electricity less expensively. We have not yet been able to harness fusion reactions, but it has been calculated that the amount of tritium released from a hypothetical fusion reactor would be 20,000 times the amount released by the generation of an equivalent amount of electricity by a fission reactor. It has been stated that release of the tritium generated by a power economy, if nuclear power were all fusion, would result in unacceptable worldwide dosages by the year 2000.[11]

We need to continue to be on our guard against damage to our environment by radiation. Of course we cannot eliminate background radiation, and we need some of the tools and techniques which employ radioactive processes, but we need to be sure that the amount of radiation reaching the environment is minimal and that the benefits outweigh the long-run cost.

NOTES TO CHAPTER 8

1. Clarence T. Lange, Ray R. Lemmerman, and Glenn Farrell, "Introduction to Radiobiology," *American Biology Teacher,* 27 (1965): 421.
2. Ibid., 423.
3. Robert W. Holcomb, "Radiation Risk: A Scientific Problem?" *Science,* 167 (1970): 853.
4. Lange, 424.
5. Kathleen Sperry, "Radiation Hazards: Senate Bill Would Provide Federal Regulation," *Science,* 157 (1967): 1292.
6. Robert W. Miller, "Delayed Radiation Effects in Atomic Bomb Survivors," *Science,* 166 (1969): 573.

7. Ibid., 570.
8. E. J. Sternglass, "Cancer: Relation of Prenatal Radiation to Development of the Disease in Childhood," *Science,* 140 (1963): 1102.
9. Ibid., 1102 f.
10. Eric Reiss, Irene Walter Johnson, Malcolm Peterson, and Virginia Brodine, "Environmental Radiation Hazards," *American Biology Teacher,* 27 (1965): 496.
11. Frank L. Parker, "Radioactive Wastes from Fusion Reactors," *Science,* 159 (1968): 83.

Chapter 9

Pesticides and Herbicides

Another area of environmental damage is that brought about by the introduction of pesticides and herbicides into the environment. Pesticides are synthetic substances which kill insects, and herbicides are synthetic chemicals which destroy weeds by interfering with their metabolic processes. DDT, the best known of the pesticides, was first synthesized in 1874 by Zeidler when he was working on his Ph. D. at Strasbourg. Later it was patented by a Swiss firm.

In 1915 Bayer, a German firm interested in dyestuffs, began looking for material to mothproof wool. They found that when a yellow substance was added to the dye some mothproofing was effected. This was the basis for studies which led to various mothproofing substances and suggested the use of DDT. The first introduction of DDT was in a Swiss compound known as Gesarol.

Other pesticides include dieldrin, aldrin, endrin, heptachlor, chlordane, and lindane. Since the introduction of DDT as an insecticide in 1939 its use has skyrocketed until very recently. Between 1957 and 1967 production ranged between 99 and 179 million pounds per year. In 1967 production was about 103 million pounds. If U. S. production is 75 percent of the world production, we may assume that the world equilibrium is based on an annual release of about 200 million pounds into the biosphere. The total amount of DDT residue eventually circulating in the biosphere would be about 3 billion tons, but this equilibrium would be approached only after 75 years. If we had used DDT at this recent rate since 1946 we would now have 1.5 billion pounds in the biosphere or about one half the total residual equilibrium we would have under these conditions.[1]

It is estimated that the half-life of DDT residues in biological systems is at least 10 years. This figure is a minimum; the persistence may be even greater. Because the chlorinated hydrocarbons are synthetic substances, they are not broken down by bacteria and fungi in the soil. It is for this reason that they tend to accumulate.

Recently, largely as the result of the appearance of Rachel Carson's *Silent Spring* in 1962 and a consequent increasing concern for the introduction of DDT into the biosphere, the quantity of DDT produced has dropped markedly.

DDT has made a substantial contribution toward saving lives and toward increasing our crop supplies during World War II and following. It has proved unmatched in the worldwide battle against such killers as typhus, encephalitis, and particularly malaria. There is no doubt that its mastery over the mosquitoes that carry malaria has spared millions of people from death and debilitating infection.

Equally important has been its effect on crop yields. The yield of U. S. cotton fields has almost doubled in the past two decades because it has been possible to control the boll weevil through the use of DDT. The National Agricultural Association believes that the use of DDT has resulted in increased yields worth between $5 and $100 an acre in such crops as barley, tomatoes, sugar beets, and cottonseed.[2]

Upsetting Natural Balances

Unfortunately DDT and other pesticides have seriously upset the balance of nature in many instances. Farmers spraying their orchards to get rid of insect larvae which consumed the fruit inadvertently killed a great many honey bees and in some cases actually reduced their crop yields. An experimental spraying to combat the spruce budworm in Montana resulted in a marked reduction in the fishing in the Yellowstone River.

The best recent example of how DDT can upset the balance of nature is what happened in Borneo after the World Health Organization sprayed huge amounts of the pesticide there. The pesticide killed the house flies, and these were in turn feasted on by the geckos or lizards. The geckos in turn were devoured by local cats. Unhappily the cats died in such large numbers from this diet that the rats they once kept in check began to overrun whole villages. Because of the threat of bubonic plague WHO officials were forced to replenish Borneo's supply of cats by parachute.

Organisms other than insects have been killed. In Florida an estimated 1,117,000 fish of at least 30 species were killed when dieldrin was sprayed to kill the sand flies. Crustaceans were virtually exterminated; similarly, fiddler crabs survived only in areas missed by the treatment.[3]

In 1963 there was a "silent spring" in Hanover, New Hampshire. Seventy percent of the robin population—between 350 and 400 robins—was eliminated by spraying for dutch elm disease with 1.9 pounds of DDT per acre.

Similar instances were recorded on the campuses of Michigan State University and the University of Wisconsin. At the University of Wisconsin the substitution of methoxychlor has decreased bird mortality; the robin population jumped from 3 to 29 in a 61-acre area following a change from DDT to methoxychlor.

One of the classic examples involving the widespread destruction of nontarget organisms was the fire ant eradication program in our southern states. In 1957 dieldrin and heptachlor were aerially sprayed over 2.5 million acres. Wide elimination of vertebrate populations resulted, and recovery of some populations is still uncertain. A subsequent study by the Georgia Academy of Science reported, ironically enough, that the fire ant is not really a significant economic pest but a mere nuisance.[4]

Pesticides Accumulate

One of the problems with pesticides is their gradual accumulation and their increase at each trophic level. A 46,000-acre warm lake in California, Clear Lake, north of San Francisco, was sprayed for gnats in 1949, 1954, and 1957 with DDD, a chemical believed to be less toxic than DDT. Later it was found that the plankton of the area contained 265 times more of the chemical than originally applied, the frogs 2,000 times more, the bluegills 12,500 times more, and the grebes up to 80,000 times more. In 1954 death among the grebes was widespread. Prior to the spraying at least a thousand of these nested on the lake. Subsequently no grebes hatched for 10 years.[5]

A major concern is the effect that DDT is having on our birds. There is considerable evidence that it is responsible for the decline of some of our birds such as the robin; these have been the victims of a futile attempt to save our elm trees by spraying them with DDT.

Even when birds are not killed by DDT some bird species may suffer irreparable damage, for there is considerable evidence to suggest a lowered reproductive potential when the pesticides occur in the egg in sufficient quantities either to prevent hatching or to decrease vigor among the young birds hatched. Birds of prey such as the bald eagle, the sparrow hawk, and others are in serious trouble.

During the years 1950 – 1965 a population crash of peregrine falcons occurred in North America and in parts of Europe and it may be that the peregrine is faced with extirpation over much of its former range if not with actual extinction.

One drastic change that seemingly affects several populations of birds that show high concentrations of insecticides is a reduction in eggshell thickness that results in egg breakage and egg eating by the parents. This is associated with the general failure to lay eggs, a decrease in the number of eggs, a disinclination to renest, and reduced viability of the young. The timing of these events coincides with the general introduction of DDT or its metabolites, particularly DDE.[6]

Similar declines are reported for the Bermuda petrel, for sparrow hawks, for the bald eagle, and for other raptors that are at the end of long food chains. Predators at the end of short food chains are not affected.[7]

Tests have revealed unexpectedly high concentrations of DDT and its residues in Canada's polar bears. These are at the top of a food pyramid, and it is possible that pesticide residues may eventually reach even higher levels in this species. Polar bears are of considerable economic importance to Eskimo and Indian hunters.[8]

Another problem in connection with the use of insecticides is that insects have a remarkable ability to develop a resistance to insecticides. Flies have developed a resistance to DDT, and resistance has appeared in the codling moth on apples and on certain cotton, cabbage, and potato insects. Over 100 important insect pests now show definite resistance to insecticides.[9]

Also to be considered is that the interaction of two compounds may result in a third much more toxic than either one alone. Malathion is considered a relatively "safe" insecticide because it is broken down by the liver and does not accumulate in human beings and animals. However there are substances which may interrupt the enzyme system which breaks down malathion, and the combination of these and malathion may be exceedingly toxic.

After two decades of intensive use pesticides are now found throughout the world, even in places far from the actual spraying. We have referred to the situation among the Canadian polar bears. Penguins and crab-eating seals in Antarctica are contaminated, and fish far from the coasts now contain insecticides ranging from 1 to 300 parts per million in their fatty tissues.

The major rivers of our nation are contaminated by a variety of chlorinated hydrocarbons. Endrin reached its maximum in the lower Mississippi in the fall of 1963 when an extensive fish kill occurred.[10]

Since chlorinated hydrocarbons are not broken down, they may accumulate in a given ecosystem in mud and in water; they need not accumulate only in living things. This may sterilize these materials and create a major problem extending far into the future.

Just how dangerous are these DDT residues so far as man is concerned? We must admit that we do not know. We have no clear evidence that DDT or its breakdown products such as DDE harm human beings, but it is unlikely that a substance causing extensive reproductive damage to birds and known to be toxic in other organisms is totally harmless to man. One writer says "The peregrine falcon may be a warning to mankind. There is little direct evidence today that pesticide deposits in man are harmful but the effects (as demonstrated in other animals) are so difficult to detect, so insidious, that they may take generations to discover. Man can gamble that he is much more resistant to chlorinated hydrocarbons and other environmental pollutants than animals which have so far shown effects. The stakes are high." [11]

The Food and Drug Administration has set a maximum tolerance level of 5 parts per million of DDT in food products. In 1968 the Food and Drug Administration seized 10 tons of coho salmon from Lake Michigan because they contained 19 parts per million of DDT. Studies indicate that the average human being has 12 parts per million in his tissues.

Mothers' milk contains more than the recommended daily intake of DDT, so the chairman of the committee studying pesticides for the Swedish National Research Council advocated that mothers switch to feeding cow's milk because cows secrete only from 2 to 10 percent of the DDT they ingest.[12]

Herbicides

Herbicides are chemical weed killers used to control or kill unwanted plants. Following World War II the chlorinated herbicide 2, 4-D began to be used widely on broad-leaved weeds. Later 2, 4, 5-T was added. This proved to be especially effective on woody species. Today over 40 weed killers are available. Most of these are used in agriculture, but considerable quantities are also used for aquatic weed control and in forestry and wild life and right-of-way vegetation management. Quantities are also being used as defoliants in Vietnam.

Although herbicides in general are much safer than insecticides, there is great danger in their indiscriminate use which results in much habitat destruction. Sprayed herbicides may drift with the air and be ingested by animals and by man; these materials may then have an adverse effect.

One problem of 2, 4-D is that some poisonous plants normally unpalatable to livestock become palatable when they are sprayed with 2, 4-D. If these are then eaten the effects may be disastrous. In addition certain palatable plant species are made toxic because of the excessive nitrate accumulation.[13]

Recently the use of 2, 4, 5-T has been curtailed since there is strong evidence that it and other herbicides cause birth malformations in animals.

A Yale biologist, Arthur W. Galston, believes that there is a possibility that the use of herbicides in Vietnam is causing birth malformations among infants of exposed mothers. He believes that though laboratory tests do not prove 2, 4, 5-T and 2, 4-D are able to cause birth malformations in humans at the dose levels experienced in Vietnam, tests do suggest the possibility. While individual exposure to these chemicals in the United States is lower than in Vietnam, it too may be a hazard here.[14]

Other problems have arisen in connection with chemical treatment. Montana's Department of Health and the State Fish and Game Department warned hunters that game birds in the state contain more mercury than humans can tolerate. The high mercury content is believed to have come from the organic mercury fungicides used to treat grain. Mercury content in the birds ranged from 0.05 to 0.47 parts per million; the tolerance level suggested by the World Health Organization is 0.05 parts per million.[15]

Other reports of toxic amounts of mercury in our soil and waters are beginning to accumulate.

There is no doubt that both herbicides and pesticides have their place. DDT has done much good in controlling disease. Pesticides and herbicides have made it possible to increase our crop production, and it is generally agreed that without these the cost of food production will increase markedly.

Indiscriminate Spraying

What is to be regretted has been the widespread, indiscriminate use of both herbicides and pesticides. No attempt has been made to limit their application, and it is because of this that larger quantities than are desirable have been used and have accumulated. The use of herbicides and pesticides is now being restricted by law. California bans the use of DDT in homes, gardens, and in crop dusting.[16] The Department of Agriculture has reduced the use of pesticides in its program and substituted less toxic, less persistent pesticides. Because of the indiscriminate use of these substances, we shall probably have to forego their use entirely.

Biological Control

Probably we need more emphasis on biological control. The first example of successful biological control was the control of the cottony-cushion scale insect which threatened to wipe out the citrus industry in California. It was brought under control by the introduction of the Vedalia beetle. The European corn borer, the spotted alfalfa aphid, the alfalfa weevil, and the woolly apple aphid have been controlled by introducing their natural enemies.

The Tokelau Islands in the South Pacific have recently been the scene of trials of new techniques in the control of mosquitoes. A fungus pathogenic to mosquito larvae has been introduced. Some 2,000 Argentina flea beetles are being used against the channel-choking alligator weed. This tiny insect feeds entirely on this aquatic weed, which is a serious impediment to navigation and stream flow in 3,000 miles of inland waterway in the southeastern states from North Carolina to Texas.[17]

A large saltwater snail has been introduced to eat submerged weeds in the southern United States and in Puerto Rico.

Birds are important in the control of both insects and weeds. Many birds feed exclusively on insect larvae; others consume weed seeds. Indeed, one of the tragedies of the use of DDT has been the fact that the killing of birds by DDT has aggravated the insect problem.

Other Methods of Insect Control

Another technique to reduce damage by insects is sterilization of male insects by gamma rays and their release into wild populations. This is effective where the population is low, but this is also particularly effective in the case of insects which mate only once. This technique has been used quite successfully against the screw worm.

Still other methods of insect control involve the use of synthetic hormones which prolong the larval period and prevent the reproduction of insects. A number of such studies are under way.

We need more research in the area. We need to know exactly what the hazards of pesticides and herbicides are, and we need to study ways in which other techniques may be applied to control insect pests and plant pests. The "balance of nature" concept would suggest that we employ biological control methods wherever possible.

NOTES TO CHAPTER 9

1. George M. Woodwell, "Radioactivity and Fallout: the Model Pollution," *BioScience,* 19 (1969): 886.
2. William A. Niering, "The Effects of Pesticides," *BioScience,* 18 (1968): 869.
3. Ibid., 870.
4. Ibid.
5. Clarence Cottam, "The Ecologists' Role in Problems of Pesticide Pollution," *BioScience,* 15 (1965): 459.
6. George H. Lowery Jr. "Peregrine Falcon Populations," *Science,* 166 (1969): 591.

7. *BioScience,* 20 (1970): 60.
8. *The American Biology Teacher,* 32 (1970): 38.
9. Niering, 871.
10. Ibid.
11. *BioScience,* 20 (1970): 60.
12. Beryl L. Crowe, "The Tragedy of the Commons Revisited," *Science,* 166 (1969): 1107.
13. Meyer Chessin, "Controversial Uses of Herbicides," *Science,* 166 (1969): 310.
14. Bryce Nelson, "Herbicides: Order on 2-4-5T Issued at Unusually High Level," *Science,* 166 (1969): 977.
15. *Science,* 166 (1969): 976.
16. *Science,* 165 (1969): 677.
17. *American Biology Teacher,* 31 (1969): 584-6.

Chapter 10

What Is Needed

How are we to solve our environmental problems? What can we do to arrest deterioration of the environment and to correct the abuses which we have inflicted on it?

Present Federal Legislation

There is no doubt that we need legislation. The Water Quality Act of 1965 and the Air Quality Act of 1967 are steps in the right direction. Congress in 1969 passed the Environmental Policy Act, which is another step forward.

The Environmental Policy Act

While it is true that this act amounts to no more than a statement of good intentions it is at least a step forward. The act has two major features. The first consists of a declaration of policy that is made more meaningful by "action forcing" provisions, prescribing specific procedures to be followed by federal agencies as they develop policies and plans which affect the environment. The second requires the president to submit to Congress an annual environmental quality report and to establish as a part of the executive office of the president a high-level council on environmental quality. Congress will hold hearings on the presidential report which the new council will have the task of preparing.

The Environmental Policy act is loosely analogous to the Employment Act of 1946. It calls on the government to seek environmental enhancement by all practical means consistent with other essential considerations of national policy. The policy goals include having an environment supporting diversity and individual choice, obtaining to the maximum extent possible the recycling of depletable natural resources, and achieving a balance between population and resource use which will permit high standards of living and a wide sharing of life's amenities.

According to the act each person "should have a healthful environment" and has a "responsibility to contribute to the preservation and enhancement of the environment." The Senate version stated "each person has a fundamental and inalienable right to a healthful environment," but this language was deleted in conference at the insistence of the House conferees.

The Need for New Laws

Regulatory laws are needed both on the federal and state level to protect the commons from abuse. Individuals who seek to pollute their surroundings will have to be penalized. In addition we shall probably want to require people to pay the costs of cleaning up whatever they discharge into the air, water, or soil. Some have even proposed solving pollution problems not by prohibiting the discharge of materials into air and water but by taxing heavily those who persist in doing this. However this smacks of a "tax on sin" and suggests that money will solve our environmental problems. It would seem that a better procedure is to forbid the discharge of wastes into the environment.

In addition to regulatory agencies we need management agencies. Such governmental agencies can carry on research and make suggestions as to how industries, municipalities, and individuals can avoid polluting the environment. We need a great deal of such research to find practical solutions to many of our problems.

Research Needed

The fact of the matter is that in many cases we still do not understand the extent of damage to the environment or the effects on human health and welfare of environmental pollutants. There is a real danger that we will be carried away by rhetoric unless we have the factual support for our statements that research provides. It is indeed true that extrapolation from animal research to human beings is hazardous, and it is quite possible that the effects we see in animals will not develop in man. Conservation suffered in its early years because in their enthusiasm some conservationists became prophets of doom, and when their predictions did not come to pass they were discredited. There is real danger of a similar "credibility gap" today. We face real problems, but it would be tragic if the hazards were ignored because of statements of well-meaning and enthusiastic but misguided conservationists.

So many of our statements at present are mere "guesstimates." What is needed first is a collecting of facts and data. In many cases the data which we have are actually inadequate. It is to be hoped that we will find it possible to spend substantial sums for data-gathering research.

We also need to avoid scapegoating. At the present time science and technology are under considerable criticism. Many people believe that these are responsible for our environmental problems. "Given the actual disasters that scientific technology has produced, superstitious respect for the wizards has become tinged with the lust to tear them limb from limb" and is expressed as "murderousness toward scientists as persons, more like anti-Semitism."[1]

In this connection we need also to avoid a romanticizing of nature. Few of us would be willing to return to a natural situation. Many of the alterations in our environment have indeed been beneficial. Nancy Ayres writes: "This is not to condemn technical, industrial, and urban developments as such. Things are not as they 'used to be', but then they never were, and nature shouldn't be romanticized either. But we are also faced with real and unnecessary losses if we do not use our best judgment with imagination and restraint, both to preserve and to develop a significant range of natural resources.

"Conservation requires planning for new possibilities as well as planning to conserve what is already here. The way our cities grow—or sprawl—affects the rest of our land, water, plants, and animals. And how we use our water, our mountains, plains, rivers, streams, ponds, lakes, subsurface water, subsoil, trees, shrubs, flowers, insects, birds, and the air itself affects us all. *These are the roots of our humanity.*

"One hidden peril of an industrial and urban culture is that so much responsibility seems remote and impersonal. So much seems to be decided for us. So many materials and tools seem to consist exclusively of paper and words that many of our sensitivities, emotional responses, skills, and judgments atrophy."[2]

In addition to facts and data we need research into practical ways to solve our pollution problems. We have referred earlier to the desirability not only of government regulatory bodies but also of "management" bodies to suggest solutions to our problems. Industry should also get into the act through the support of pollution research organizations that do for pollution what organizations such as Battelle Memorial Institute and Mellon Institute do for other research problems. The recycling of pollutants is a real challenge to American industry. They supply a potentially usable raw material. We've learned to utilize many of the materials once discarded; the meat-packing industry is said to use everything except the pig's squeal. It should be possible to recycle and use the materials which we discard to pollute our environment.

Needed: An Ethic

There is no doubt that we need a scientific ethic. Science and technology have raised problems in a variety of areas: international relations, medicine, population and, of course, the environment. In commenting on our need for an international relations ethic, one writer recently said: "In the past man's basic problems were concerned with the complexities of his physical environment. To survive and flourish it was necessary for him to learn how to deal with the elements, to provide adequate food and water supplies, to master the challenges of distance and communication, and to acquire sources of energy beyond those available through the use of his own musculo-skeletal system. It is clear that in our time, the focus of the 'basic problem' has changed from the external to the internal environment. Our primary need is no longer one of coping with the physical universe; it is now instead a question of learning and learning rapidly how to cope with ourselves, with each other, and in particular with intergroup and international conflicts. This is the fundamentally new condition in response to which we must hope that we will be capable of rational and anticipatory changes. Our need is for a new Manhattan project devoted not to the development of a weapon, but instead, to the development of a new body of knowledge of intergroup relationships and conflict resolution so that we can preserve freedom and peace."[3] What he has said applies also to environmental problems: we need to learn how to use the resources God has provided, resources which modern science has unlocked and put at our disposal.

Garret Hardin recently raised the question: "Who is the specialist in moral questions? As nearly as I can make out it is somebody who bears either the name of a philosopher or a theologian. When I look at the work the philosophers and theologians do on ethical questions I lose all diffidence about entering this area. It seems to me that the gravest shortcoming of these people is that they operate in a way that is essentially nonproductive. When they tackle a moral problem it is with a way that indicates they really cherish the insolubility. The moment they even come within helping distance of solving a problem, they immediately discover all sorts of objections why this can't possibly be done and run the other way.

"I think this is a deeply engrained attitude of mind of theirs, that these things can't be solved and they are jolly well going to prove to you that they can't be solved. I think it was C. P. Snow who remarked that scientists 'have the future in their bones.' This is one of the striking differences between scientists and other people. A scientist or engineer attacking a problem assumes as a matter of course that it has a solution. This is entirely different from the person who says 'Of course it is insoluble, so now let me anguish over it for two or three hours.' You don't get any place that way."[4]

Actually it is unlikely that science will be able to provide the guidance he proposes. Science lacks the standards God's Word provides—standards which we shall discuss shortly. It has only one criterion for the good, and that is "Does it work?" Such a criterion leaves a great deal of room for the man who argues he can profit by exploiting the environment at the expense of others and of future generations.

However Hardin does call attention to a weakness of many modern theologians. Anxious to make their peace with science, they are willing to bend theology and the Scriptures in whatever direction the winds of science seem to be blowing. Moreover, many of them have abandoned as hopelessly old-fashioned an authoritative Word; thus they have no firm basis on which to proclaim God's will.

Is Christianity to Blame?

There are many who believe that we need a humanistic environmental ethic. They blame the Christian church for our present crisis. Lynn White Jr., in a widely quoted article, places the blame squarely on the shoulders of the Christian church. Christianity, he says, is the most anthropocentric religion the world has ever seen. "Christianity in absolute contrast to ancient paganism and Asia's religions (except perhaps for Zoroastrianism) not only established a dualism of man and nature but also insisted that it is God's will that man exploit nature for his proper ends . . . . By destroying pagan animism Christianity made it possible to exploit nature with a mood of indifference to the feelings of natural objects . . . . Somewhat over a century ago science and technology joined to give man powers which to judge by many of the ecological effects are out of control. If so, Christianity bears a huge burden of guilt . . . . We shall continue to have a worsening ecological crisis until we reject the Christian axiom that nature has no reason for existence except to serve man." [5]

White is probably correct in blaming the crisis on exploitation. He may even be right in blaming Christians for exploiting though they are certainly no guiltier than those who do not share their faith. But he is wrong in fixing the responsibility for encouraging this exploitation. Like so many other people he has forgotten to read beyond the first pages of the Bible. God's command to man to subdue the earth and to have dominion over the fish of the sea and the birds of the air and over every living thing that moves upon the earth can only be understood against the background of the Biblical concept of God's ownership of all earthly resources and man's position as a steward of what God has committed to him. It is true of course that some have quoted these verses from the first chapter of Genesis to justify exploitation of the environment. But we must remember that the Bible has been the object of frequent misquotation; there have been some who have quoted the curse on Canaan to justify racism. Even Satan misquoted the Scriptures when he tempted our Lord.

Man's Superiority

Man is indeed the foremost of the visible creatures; this is clear from Scripture. Moreover, he is to subdue the earth and have dominion over it; for this purpose God gave him a superior brain and set him apart from the animals by endowing him with the ability to communicate so that the culture and learning of the past can be transmitted to future generations. Yet man's assignment to rule over the earth and subdue it must be read in its context and against the background of other Biblical concepts.

God Owns All

Man can hardly claim to own anything. The Psalmist proclaims "The earth is the Lord's and the fullness thereof" (Ps. 24:1), and God says through the Psalmist "Every beast of the field is Mine and the cattle on a thousand hills. I know all the birds of the air and all that moves in the field is Mine." (Ps. 50:10-11)

Throughout Scripture God is pictured as the Creator: the land, the plants, the animals, the air, the water are His because He made them. Man stands in a creature relationship to God; all that he has comes from the God who created him, too. Thus man cannot claim to own anything. He hardly can claim the right to exploit.

Old Testament Property Laws

Especially interesting in this connection are the Jewish property laws of the Old Testament. The Promised Land was given by God to His people; they possessed it not as individuals but as a nation. The land was assigned to God's people by tribes, and care was taken to guarantee that the land would remain in the tribe. An individual might not transfer a title outside his tribe, Num. 36:5-9. Moreover, there was no such thing as selling the land; at best under the Jewish theocracy an individual could lease a piece of property, since in the year of jubilee it returned to its original owner, Lev. 25:13-17.

Responsible Stewardship

To these Old Testament land ownership concepts, which existed under the theocracy that God Himself established, must be added the concept of responsible stewardship which runs throughout the pages of Scripture. Because man is the crown of God's creation and because he has been given great intellectual endowments, man has a special responsibility and is expected to care for what has been entrusted to him. He does not possess it; it has been given him to husband and tend, just as our first parents were to till the garden and keep it.

God's Concern for Creatures Other Than Man

God is concerned also for the plants and animals which He created. He clothes the lilies of the field in splendor greater than Solomon's, and He cares for the insignificant sparrows, not one of which falls to the ground without His knowledge and permission. If God is concerned for other living things, man, His steward, must show similar concern.

Materialism

One of the reasons for our environmental crisis is the crass materialism of our day. Men have exploited our natural resources in order to pile up profits for themselves. They have damaged the "commons," seeking to avoid the cost of the damage which they have inflicted, in order to pile up greater and greater profits. Here too the Bible has something to say about those who consider "things" all important and who plan to build larger barns to accommodate all their possessions. Our Savior devoted a large part of the Sermon on the Mount to pointing out the foolishness of material cares and concerns. A Christian steward takes the long-range view of that which has been committed to his charge. He realizes that he cannot pile up short-time profits for his own benefit—at the expense of that which he will one day be obliged to turn over to his successors.

Man's Obligation to Other Species

Does man have an obligation to other species? May he exterminate a form which threatens his welfare or his prosperity either directly or indirectly? As we have pointed out God certainly cares for species other than man. We live in a species-poor world and while it can be argued that few people miss the passenger pigeon today and few are likely to miss the whooping crane, the California condor, and the large mammals of Africa if we exterminate them, man can hardly deny his stewardship responsibility in protecting also this aspect of God's creation over which he must rule.

At times man does face real problems in his relationship to other species. Recently a problem has arisen in our national parks in connection with the grizzly bear. These bears may well be a highly dangerous species to man, and there have been those such as Gairdner Moment who have argued that they need to be eliminated from our national parks, though they believe we ought to try to preserve these species in special grizzly bear sanctuaries. It would seem that man ought to be able to set aside some of the land to preserve these species which do constitute a threat to his welfare, and it ought to be understood man enters such areas at his own risk. This might be done in certain national parks or in specially designated preserves from which man is ordinarily excluded.

Another problem is that of competition of animals with man for food. The bison of our prairies competed with the cattle of settlers, and in most cases the competition was resolved in favor of cattle. Today the large mammals of Africa are competing with natives for food. An argument can be made for the fact that they are consuming grass which the natives need for their cattle or that the land they occupy needs to be fenced so that agriculture can expand to meet the current needs of these developing nations.

Moreover it is hard to punish a poacher who brings down one of these large animals in order to feed his family.

Yet it would seem desirable to set aside some areas where these organisms can be protected and can survive. These areas will need to be substantial, and they ought to be set aside in perpetuity to keep these organisms from extinction.

Similar areas are needed to protect some of the unusual plants. In many cases their range is extremely limited, and often they cannot be transplanted to another habitat. Indeed in many cases there are probably unidentified species of unknown value to man that are threatened with extinction.

Economic Conflict

Still another problem arises when the protection of threatened or presumably threatened species interferes with man's economic needs. One reason for the near extinction of the bison was that these animals were a threat to the expansion of the railroads following the Civil War. It sometimes took several days for the large herds to cross the tracks of the first transcontinental railroad and while they were crossing the trains had to halt. The railroads actually hired buffalo hunters, of whom the best known was Buffalo Bill, to kill the bison so that the trains might move.

Another example of the clash between man's economic interest and plans are the controversies going on in redwood areas of California. A number of conservationists have mounted campaigns to set aside the remaining redwood groves so that they will not be timbered. Yet these trees are important in the economy of the area, since they are a significant lumber source. Timber companies that lose access to the groves will have to shut down. Not only will the companies be affected, but those who depend on them for employment will be adversely affected.

The question here is just how much of our redwood reserves need to be preserved in perpetuity. Few people would argue that we need to preserve *all* of the redwoods just for the sake of preserving them (though we would hardly agree that once you've seen one redwood you've seen them all), but there is a real possibility that if we do not set aside large enough areas the redwoods will become extinct.

Still another example of this problem is the controversy surrounding the use of the southern tip of Lake Michigan east of Chicago. This area, the Indiana dunes, is a unique area from a botanical standpoint. It is a place where the ranges of a number of plant species overlap; many southern species reach the northmost point in their range here, and many northern species have the southernmost point of their range here. Because it is close to a metropolitan area and therefore lends itself to industrialization, the dunes area is of considerable importance economically. Moreover, being adjacent to Lake Michigan, the area has a potential for setting up a port which will serve the surrounding area and make possible cheaper transportation of goods.

Tensions such as these are likely to continue to arise. They need to be settled on the basis of principles set up in advance—away from the emotional heat which such controversies inevitably generate when they arise.

The Whole Creation Groaning and Travailing

The Christian cannot help but be impressed with the fact that our environmental crisis is a good example of how the whole creation groans and travails because of man's sin, Rom. 8:22. The basic reason for exploitation is man's selfishness, his refusal to practice love to his fellowman and to other creatures, and his desire to serve his own personal ends. So ruthlessly man inflicts scars on the environment. Thus because man is a sinner, the whole of creation must groan and travail because of his abuses.

God's World: A Good World

The world which God created was a good world; this is evident when one surveys his surroundings. The balance of nature is good. Each organism has its place in the ecosystem. Of course the balance is not perfect, but into the system He created, God has incorporated the ability to correct some of the imbalances which gradually arise. It is man's arrogance and selfishness that inflict on the natural world upsets greater than nature is able to cope with. Man needs to see himself as the destroying biotype.

Of course no one would argue that man ought to return to a "natural situation." As civilization has progressed we have developed greater and greater needs. The resources which God created are there for our use; it is the abuse of these resources and man's exploitation of them at the expense of fellowman and of other species that is wrong. To argue that man ought to go back to a natural life form is to ignore man's assignment to rule over the earth and subdue it. Development of the energy sources which God has provided—the coal, the natural gas, the petroleum, and atomic power—and his use of these for man's own welfare is certainly not wrong; nor is it wrong to use the mineral resources and the timber. What is wrong is selfish exploitation. Those who would argue that man ought to give up hunting, that he ought not use any of the world's resources, and that he ought cut no trees are making extreme statements; and in their enthusiasm they are likely to harm the very program which they are seeking to support. There is nothing wrong, for instance, with deer hunting; it is apparent that in many cases if the excess deer are not taken by human hunters they would die through disease. Nor is there anything wrong with harvesting ripe timber; it will replace itself in the natural course of events. It is this sort of use of the earth's resources that God had in mind when He spoke of man's ruling over the earth and subduing it.

The Need for Research

It is rather easy to state principles, but it is not easy to apply the principle of responsible stewardship which we have outlined above. This underscores the need for more and more research, an important part of ruling the earth and subduing it. Some of the exploitation of the environment has taken place because man has not recognized the damage which he was inflicting on his surroundings and has not realized the potential hazards to his own health. We need a great deal more research. We need to gather facts; we need more statistical data on our exact environmental situation. Then we also need additional research so that we may understand the complexities of the ecosystem of which we are a part and then understand at what point damage may result.

If we carry on additional research we may be able to resolve some of our differences. For instance just how much of the area of the redwoods in California needs to be set aside to protect this species? How much of our redwood forests can we cut without endangering the survival of these magnificent trees? How much of a buffer area do we need, that is, an area which is not necessarily covered by the trees themselves? How are these trees affected by air pollution? What are the effects on the groves of other alterations man is making in the environment?

At the present time in many cases we have far more questions than we have answers, and in some cases we do not even have enough information to ask the proper questions. Persons who are interested in practicing responsible stewardship will want to support every effort to gather the needed information. We need the Christian environmental ethic outlined in the Bible, and we need Christians to practice it.

NOTES TO CHAPTER 10

1. Harvey Brooks, "Physics and Polity," *Science,* 160 (1968), 397.
2. Nancy Ayres, "The Roots of Our Humanities," *Nature Study,* 22 [2] (Summer, 1968), 4 f.
3. William Pollin, Book Review of Fromm's "May Man Prevail," *Science,* 135 (1962): 306. (Copyright 1962 by AAAS.)
4. "Biology in the Next Two Decades," *Commission on Undergraduate Education in the Biological Sciences News* 6 [1] (October 1969), p. 4.
5. Lynn White, Jr. "The Historical Roots of Our Ecological Crisis," *Science,* 155 (1967), 1203-7.

Glossary

By Andrew J. Buehner

The ecological emphasis of our times has brought with it a new vocabulary or a new and popular use of words and phrases from various disciplines. This glossary is offered as a guide to readers in this period of ecological crisis.

alewife – small North American fish of the herring family.

aquifer(s) – the underground water table.

autophyte – a plant which synthesizes its food from inorganic substances.

biodegradable – substance capable of being broken down by bacteria and fungi in the soil or by other environmental processes.

carnivores – flesh-eating organisms.

climax predator – an organism preying on others and affecting the equilibrium of the community.

detergent – a mixture of compounds for use in washing which can emulsify oils and hold dirt in suspension; however, they contribute to eutrophication of lakes and rivers.

DNA – deoxyribonucleic acid.

decibel [db.] – a measure of the intensity of sound; it ranges from the least audible sound for humans (0-1 db.) to a maximum of about 130 db. (Eardrums may rupture at 140 db.) Continued exposure to 100 db. may be deafening. A hard rock band may peak at 120 db.

ecosystem – a self-sufficient unit or system in nature with simple or complex cycles. It involves living organisms and nonliving matter.

emphysema – a lung condition of air-filled expansion of body tissues; heart action often impaired.

eutrophication – nutrients, especially detergents, entering water, speed up the natural aging process by causing excessive growth of algae and other aquatic plants which decompose and use up the water oxygen vital for fish and animal life. Besides detergents, man-made fertilizers and human and animal wastes help accelerate the process.

filter feeder – an organism which obtains nutrients by filtering a current of water passing through a part of its system.

herbicides – plant or weed killers that interfere with normal growth-life.

herbivores – plant-eating animals.

heterophytes – plants that get their food materials from other living or dead plants or animals or their products. They decompose complex materials into simple organic compounds.

in utero – in the womb.

keratin – sulphur containing fibrous proteins which are the chemical basis of epidermal tissues.

laterization – a rocklike hardening of tropical soils containing aluminum and iron hydroxide.

monoculture – the cultivation of a certain product (e.g., corn or wheat) to the exclusion of other possible products or uses of the land.

myxomatosis – a severe disease of rabbits characterized by fever, swelling, and inflammation, and transmitted by mosquitoes. Virus used in rabbit control.

PAN – peroxacetyl nitrate.

Penicillum notatum – a genus of fungi which in medicinal application has marked bacteriostatic effects on staphylocci, pneumococci, and other bacteria. Popular form: Penicillin.

perinatal—occurring at about the time of birth.

pesticides—substances used to kill pests, usually insects.

pollution—the defilement of a system to the point where it is unable to cleanse itself or to carry out the purposes of its creation. This applies to air, water, land, sea, natural resources—the whole environment of man and other organisms.

protoplasm—organized living matter: the whole complex of organic and inorganic substances and water which makes up a cell.

raptor—a bird of prey.

RNA—ribonucleic acid.

roentgen—the international unit of X radiation or gamma radiation.

ruminants—cud-chewing animals.

sea lamprey—an aquatic vertebrate similar to the eel but with a large suctorial mouth; very destructive of fish fauna in the Great Lakes.

sea star—starfish.

smog—originally, smoke plus fog; today, heavily contaminated air; *reducing* smogs contain sulphur dioxide (which forms sulphuric acid) plus carbon and gases; *oxidizing* smogs include gases from internal combustion engines (esp. nitrogen dioxide, ozone, and peroxacetyl nitrate).

symbiosis—a living together of dissimilar organisms for mutual benefit.

thermal inversion—a condition in the atmosphere in which cool air moves above warm air and prevents it from rising; it makes polluted warm air dangerous to life.

trophic level—the nutritional level at which an organism maintains itself.

tsetse fly—variety of African fly south of the Sahara desert; spreads *nagana* or sleeping sickness.

Index

By Andrew J. Buehner